AF564544

Reclamation and Management of Problematic Soils

About the Authors

Dr. Mahendru Kumar Gautam received his B.Sc. Agriculture degree at Acharya Narendra Deva University of Agriculture and Technology, Ayodhya, and M.Sc. Agriculture at department of soil science and agricultural chemistry, Institute of Agricultural Sciences, Banaras Hindu University, Varanasi, Uttar Pradesh and Ph.D. Agriculture at department of soil science and agricultural chemistry, C.S. Azad University of Agriculture and Technology, Kanpur Uttar Pradesh. Dr. Gautam is Presently working as Assistant professor (Soil Science), School of Agriculture and Environmental Sciences, Shobhit University, Gangoh, Uttar Pradesh.

Dr. Mahendra Singh began his journey in Soil Science during his postgraduate studies (2003–2005) at G.B. Pant University, focusing on the impact of AM fungi and Bradyrhizobium on soybean. His PhD (2005–2008) further explored the effects of manures, fertilizers, and micronutrients on soybean growth and yield. He gained industry experience from 2008 to 2012, which shaped his approach to agricultural education and research. In 2012, he joined BAU, Sabour as an Assistant Professor, guiding Master's and Doctoral students and leading research in soil microbiology. Currently, he is an Associate Professor at ANDUAT, Kumarganj, Ayodhya. Dr. Singh has published 68 research papers, 12 book chapters, and developed three commercialized liquid biofertilizer formulations recommended for farmers.

Mr. Dharmendra Kumar currently pursuing Ph.D. in Soil Science & Agricultural Chemistry from Acharya Narendra Deva University of Agriculture & Technology, Kumarganj, Ayodhya. He has received his Master’s in Soil Science and Agricultural chemistry and B.Sc. (Hons.) Agriculture from Chandra Shekhar Azad University of Agriculture & Technology, Kanpur (U.P.). He has qualified UGC NET-2023. He has published 15 research papers in different reputed journals. He has written several book chapters, Articles and a book entitled “Soil Fertility and Plant Nutrition”. He is member of different journals and reviewed many papers.

Dr. Hemant Jayant received his B.Sc. (Ag.), M.Sc. (Ag.), and Ph.D. degrees from the Department of Soil Science and Agricultural Chem istry, Institute of Agricultural Sciences, Banaras Hindu University (BHU), Varanasi. He conducted research on remediation of heavy metal contaminated soil during his master programe and during his doctoral studies, he conducted research on remote sensing and GIS mapping of soils. He has published 10 research papers in different reputed journals. He has written several book chapters, and articles. He is member of different journals and reviewed many papers. Dr. Jayant is currently serving as an Assistant Field Officer at the Soil and Land Use Survey of India, Nagpur, Maharashtra.

Reclamation and Management of Problematic Soils

Mahendru Kumar Gautam
Mahendra Singh
Dharmendra Kumar
Hemant Jayant

1168, Sector 13, Urban Estate
Karnal-132 001, Haryana
Tel: 91-84470 75807, 18440 41168
Email: contentvibesppa@gmail.com
www.contentvibes.in

PRINT ISBN: 978-93-67559-78-9

E-ISBN: 978-93-67553-73-2

Preface

All civil engineering projects start with soils, however not all soils have the best qualities for building. Engineers deal with "problematic soils" in many places of the world, which are those that present difficulties for the design, building, and long-term performance of structures because of their distinct mechanical, chemical, or physical characteristics. These soils can be very compressible, have low shear strength, swell or shrink, or collapse when wet. For infrastructure development to be safe and sustainable, these issues must be recognized and addressed. With an emphasis on problem soils, submerged soils, micronutrients, and heavy metals, this textbook on "**Reclamation and Management of Problematic Soils**" is based on my research and teaching experience in soil chemistry, fertility, and plant nutrition. incorporating arsenic into farming practices. The target audience for this book is B.Sc. (Ag). The revised ICAR syllabus serves as the basis for this. Students studying environmental sciences and related fields of agriculture, however, might also gain from it.

The book is structured into fourteen chapters, each of which focuses equally on a different topic pertaining to agricultural research in India. The chapters in this book cover the entire course and are arranged chronologically, making it simple for beginning agricultural students to understand. The treatment of various problem soil types and economical bioremediation techniques employing multiple tree species (MPTS) are covered in this article. Undergraduate agriculture students can gain from this thorough examination of flooded soils, saline irrigation water utilization, and soils influenced by acid, acid sulfate, and salt. Damaged sites and waste management concerns were also taken into account. This book focuses on using Geographic Information Systems (GIS) and Remote Sensing (RS) to monitor and assess degraded lands, irrigation water, and problematic soils. The purpose of this textbook is to meet the demands of instructors and undergraduate agriculture students at different Indian agricultural universities. Any recommendations for enhancement would be much appreciated.

My profound appreciation goes out to my coworkers, students, and the other geotechnical specialists whose knowledge and experiences have influenced the material in this book. I hope that this work will be used as a practical reference and a foundational text for anyone who deal with the particular difficulties of building on troublesome soils.

Authors

Contents

1

Introduction: Historical Perspectives

Introduction

Managing the diverse array of stresses confronting cultivated lands stands as a pressing necessity within the agricultural domain, demanding swift adaptability to navigate shifts and satisfy evolving requisites along the production-consumption continuum. In India, approximately 7.0 million hectares of agricultural terrain grapple with salt-related predicaments, a figure poised for escalation owing to factors like secondary salinization in irrigation networks, amplified dependence on substandard water reservoirs in arid and semi-arid locales, and the complexities posed by phenomena such as sea water encroachment and brackish water aquaculture in coastal regions. Forecasts indicate a potential expansion of salt-affected soil regions in India to encompass roughly 13 million hectares by the year 2025.

Within this landscape of challenges, agronomists shoulder a pivotal responsibility, entrusted with the mandate of bolstering productivity via pioneering research endeavors and the formulation of holistic strategies. This undertaking necessitates a nuanced comprehension of salt-affected soils, coupled with the crafting of contingency blueprints predicated upon resource-efficient, economically sound, and ecologically sustainable methodologies. Given the flux in climatic patterns, the adoption of proactive measures assumes heightened significance.

Scholarly discourse, exemplified by Singh's seminal work in 1998, furnishes a retrospective lens on the saga of salt-affected soils. Building upon this historical scaffold, the present exposition endeavors to trace the trajectory of agronomic inquiry spanning epochs ranging from the Vedic era through pre-independence and post-independence epochs. By spotlighting the seminal contributions of researchers in tackling salt-induced soil adversities, the paper endeavors to underscore the profound ramifications of such endeavors on societal well-being and environmental equilibrium.

Soil Salinity in Ancient India: Pre- and Post-Vedic Perspectives

Landforms and landscapes emerge from the weathering of rocks and minerals, incorporating salts as essential constituents. These salts have been present since the inception of the landscape, either in soluble forms or as components of weathered rocks and minerals constituting the soil. The accumulation of salts beyond a certain threshold is what precipitates the formation of saline lands. The earliest documented occurrences of saline lands date back to 2400 BC in the Tigris-Euphrates alluvial plains in Iraq (Russel *et al.,* 1965). These lands were initially identified and linked with irrigation practices in North-Eastern Sumer near modern-day Telloh. Their prevalence continued to escalate over the subsequent 700 years until the decline of the Sumer civilization. In antiquity, civilizations recognized the significance of watering their crops and the potential buildup of salt in the soil over time. In regions like Mesopotamia and the Indus Valley, inhabitants cultivated near rivers. They demonstrated adeptness in agriculture during both the pre-Harappan and Harappan periods. The Harappans, notably, displayed ingenuity as engineers. They constructed wells, reservoirs, and drainage systems to cultivate crops, primarily during the winter season. They inhabited areas where river floods washed away salt deposits, thus maintaining soil fertility. Subsequently, around 1500 BC, the Aryans migrated to the Indus Valley, marking the onset of the Vedic era. During this era, people began discerning lands suitable for farming and those unsuitable. They even designated names like "Usar" to denote lands affected by salt. By the middle of the second millennium AD, with increased land demand, people commenced attempts to cultivate salt-affected lands instead of solely utilizing them for animal grazing.

Pre-Independence

Moreland's documentation (1901) provides evidence of salt-affected soils in the former United Provinces of Agra and Oudh, along with insights into the traditional methods utilized by local cultivators for reclamation, indicating a pre-existing knowledge of reclamation practices in the region prior to formal scientific investigations. Historical records from the Director of Land Records and Agriculture dated July 28, 1888, indicate that the area of "usar" in the North-Western Provinces and Oudh amounted to 3,129,053 acres (1,271,977 ha), aligning with later estimates of alkali lands in Uttar Pradesh (Tiwari *et al.,* 1989). Similarly, observations before extensive canal irrigation projects in the Punjab section of the Indus plains revealed soil salinity, with a reported affected belt measuring 16 km x 4.8 km (Jameson, 1852). In response, government policy involved allocating significant areas within each village settlement for exchange with settlers encountering saline patches of land. Additionally, saline soils were prevalent in the Sind region of the Indus plains, prompting the Government of India in 1906 to seek reports from experienced local irrigation and revenue officers on the prevalence of "kallar" or alkali in Sind.

The issue of land degradation in the Western-Yamuna canal command area came to official attention in 1855 following a complaint from a farmer in Moonak village (now in Haryana, near Karnal), marking one of the earliest documented cases of soil salinity linked to canal irrigation and drawing significant interest from the British Government. Soil and water samples sent to the Royal School of Mines in England in 1865 for analysis confirmed that canal water, containing sulphates and chlorides, contributed to soil salinity through salt deposition during irrigation, redirecting future research towards understanding secondary salinization. T.E. Brown, Chemical Examiner for Punjab, had earlier suggested in 1863 that minerals rich in alkaline substances in the soil, exacerbated by irrigation water, could contribute to the issue. The impact of canal irrigation on secondary salinization was also observed in peninsular India, notably with the Nira Valley Irrigation Canal, which began operation in 1884, resulting in barren, salt-encrusted patches appearing on some of the region's best farmlands within five years.

The Reh Committee, convened in 1879, extensively explored the origins of salts in soils, highlighting the perpetual contribution of canal water to soil salinity and proposing a comprehensive scheme of experimental work primarily to be executed by the newly established Agriculture Department. This scheme, endorsed by the Reh Committee, aimed to enhance soil cultivation through salt removal, improved drainage, effective silting, deep cultivation practices, appropriate manuring techniques, and the cultivation of green crops through ploughing. Subsequent conferences resolved to initiate experiments for the reclamation of salt-infested lands, meticulously documented by Leather in 1897.

To assess the distribution of Reh across provinces, meticulous mapping was undertaken, covering 65 villages in the Akbarabad pargana of the Aligarh district, along with efforts to generate a surface drainage map for these villages. A comprehensive survey detailing the water table's depth in sub-soil waters across the entire province was also completed. Numerous agricultural experiments were launched across various locations, including Awargarh, Aligarh, and Kanpur, with observations suggesting that successful growth of Acacia in usar soil might not require any special treatment, echoing principles later advanced by the Central Soil Salinity Research Institute in Karnal.

The initial reclamation trials targeted two primary soil types: clayey soils adjacent to the Indus River and its former branches, and coarse-textured soils. Henderson, assuming responsibility in 1907, encountered barren land at a section of Mirpurkhas farm where crops failed to thrive due to high salt content, primarily dominated by sulphates. Treatment involving inundation with water and specific crop rotations ultimately rendered the land fertile. Similar efforts were undertaken at Daulatpur, targeting land with high salt content, notably chloride salts, with initial outcomes emphasizing the significance of specific crops in reclamation efforts.

Reports of soil salinity issues in canal-irrigated areas of India emerged primarily from the Eastern part of Punjab, with severe threats observed in the Western region. Traditional reclamation methods didn't universally succeed, but gypsum application proved essential in certain Sind soils. The agricultural conference at Nagpur established a committee to coordinate reclamation efforts across provinces, aiming to expedite practical results through scientific investigations.

Following recommendations from the 1903 Nira Valley Expert Committee, efforts to ameliorate salt lands in the Nira valley commenced in 1908, overseen by Mann and Tamhane. Similar waterlogging and salt accumulation issues were observed in tank-irrigated areas of Southern India, prompting experiments where various drain types were employed. Wood documented subsoil drainage work in rice lands under tank irrigation, utilizing various drain types, which remained functional over several years. Stone drains implemented in alkaline rice lands functioned effectively for five years, rendering the land fertile.

Post-independence

During the period following Independence, there was a notable surge in research dedicated to reclaiming salt-affected soils within our nation. This coincided with India's implementation of successive 5-year development plans, which placed significant emphasis on bolstering the agricultural sector. The Indian Council of Agricultural Research (ICAR) played a pivotal role by spearheading research initiatives during the second 5-year plan. These initiatives included the inception of the All India Co-ordinated Research Project on Water Management and Soil Salinity (now recognized as the All India Co-ordinated Research Project for Management of Salt-affected Soils and Use of Saline Water in Agriculture) and the establishment of the Central Soil Salinity Research Institute in Karnal in 1969.

In addition to ICAR's efforts, research on alkali soil reclamation was undertaken by various institutions including the National Botanical Research Institute in Lucknow, state-owned universities, research institutes, and agricultural departments. During the second 5-year plan, ICAR's research endeavors on land reclamation were based in Ludhiana and Allahabad. Researchers at the Ludhiana center concentrated on assessing different chemical and organic amendments, agronomic practices, and nutrient management, with their findings subsequently applied in field settings through operational research projects and demonstrations. The Allahabad center primarily delved into research on organic amendments.

The All-India Coordinated Research Project adopted a collaborative approach involving research centers spanning states with substantial salt-affected lands and brackish groundwater challenges. These centers, located in Uttar Pradesh, Rajasthan, Madhya Pradesh, Karnataka, Tamil Nadu, and Andhra Pradesh, addressed diverse aspects of soil reclamation, encompassing chemical and organic amendments, nutrient management, and crop selection. Meanwhile, research at the Central Soil Salinity Research Institute in Karnal delved into the physics and chemistry of alkali soils in the Indo-Gangetic plains, leading to advancements in soil reclamation technologies.

Founding of the Central Soil Salinity Research Institute (CSSRI)

The Central Soil Salinity Research Institute (CSSRI) was founded in Karnal under the umbrella of the Indian Council of Agricultural Research (ICAR) during the Fourth Five Year Plan, as per the recommendations of an Indo-American team on water management appointed by the Indian government in 1967. This advisory committee proposed the establishment of a national research center concentrating on issues of salinity and alkalinity in agricultural soils. Following governmental endorsement, CSSRI was initiated as a planned project in March 1969, initially situated in Hisar. However, due to the prevalence of problematic sodic soils in the Indo-Gangetic plains, it was relocated to Karnal in October of the same year.

Initially, research activities were conducted at the Karnal farm, which faced significant alkalinity issues. Recognizing the necessity for expanded research, a new farm was established at Gudha near Karnal in 1978. Moreover, in 1982, a farm was instituted at Sampla in the Rohtak district of Haryana, specifically focusing on sub-surface drainage technology in saline waterlogged soils.

CSSRI rapidly gained the confidence of the farming community through on-farm research demonstrations, outreach programs, and laboratory-to-field initiatives, attracting the attention of policymakers. By the late 1980s, CSSRI had garnered recognition from the global scientific community engaged in salinity research and forged collaborations with various international organizations.

Over the years, CSSRI has addressed numerous challenges related to salt-affected soils and their management through outreach programs and the establishment of research farms at various locations including Gudha, Sampla, Mundlana, Bhaini Majra, Bichhian, Hisar, and now Nain. Additionally, the institute operates three regional research stations in Canning Town (West

Bengal), Bharuch (Gujarat), and Lucknow (Uttar Pradesh) to tackle specific issues related to coastal salinity, salt-affected Vertisols, and alkali soils of the Indo-Gangetic plains, respectively.

Since its inception, CSSRI has been at the forefront of interdisciplinary research dedicated to the reclamation and management of salt-affected soils, evolving into a globally recognized center of excellence in salinity research.

Exploring Fundamental Studies on Saline Soil Dynamics and Characteristics

Initially, the focus of research was on comprehending the basic properties and behaviors of salt-affected soils, particularly concerning interactions between soil, water, and plants. Throughout the 1970s and 1980s, extensive studies were conducted to investigate moisture retention, water transmission, cation exchange equilibrium, and the influence of clay mineralogy on soil processes. Concurrently, there was a growing interest in understanding microbiological processes and their impact on soil microbial diversity and biological activity within salt-affected environments.

As understanding deepened, research efforts shifted towards examining solute transport processes, solute transport modeling, and hydro-salinity modeling by the 1980s. These endeavors aimed to gain comprehensive insights into solute movement dynamics and their implications for managing soil salinity.

Key research areas included

Fundamental Research on Soil-Water-Plant Relationships: Prioritizing fundamental and strategic research on the interactions between soil, water, and plants.

Moisture Retention Characteristics: Extensive studies delved into moisture retention characteristics, as evidenced by notable works such as those by Acharya and Abrol (1975).

Water Transmission Characteristics in Sodic Soils: Considerable attention was given to studying water transmission characteristics in sodic soils with varying Exchangeable Sodium Percentage (ESP) and pH, as documented by Abrol *et al.* (1978) and Acharya and Abrol (1991).

Cation Exchange Equilibrium: The importance of cation exchange equilibrium in salt-affected soils' response to reclamation technologies was recognized.

Clay Mineralogy: Emphasis was placed on understanding the role of clay mineralogy in controlling cation exchange equilibrium and various soil processes.

Microbiological Processes: Initiatives were taken to explore microbiological processes and their impact on micro-flora and fauna in salt-affected soils, including microbial diversity, biological activity, and the role of microbes in soil improvement.

Soil Fertility Management: Advances were made in soil fertility management in salt-affected soils, focusing on electrochemical changes during rice growth, nutrient dynamics and management, and the integrated use of organics with mineral fertilizers.

Solute Transport Processes: Research shifted towards understanding solute transport processes in salt-affected soils.

Solute Transport Modeling: There was a focus on developing solute transport models and hydro-salinity models to further comprehend the complexities of salt-affected soils.

Furthermore, research endeavors post-1980 increasingly concentrated on studying crop responses to salinity, aiming to find effective and economical biological interventions for salinity issues. Understanding how plants adapt to high salinity conditions became a pivotal aspect of research, with studies revealing insights into eco-physiological and biochemical adjustments various plant species make to counter salinity stress. Investigations also identified the order of ion toxicity across seeds: $NaHCO_3 > Na_2CO_3 > NaCl > CaCl_2$. Additionally, research explored environmental and genetic adaptations, focusing on natural coastal vegetation resilience to high salinity levels and salt tolerance mechanisms in plant species, which significantly influence crop responses to salt stress.

Soil salinity presents a significant worldwide challenge, detrimentally affecting agricultural sustainability and productivity. This issue transcends various climatic zones and can arise from both natural phenomena and human intervention. Typically, saline soils are prevalent in arid and semi-arid areas where limited rainfall fails to remove mineral salts from the root zone, thus hindering crop development. The historical correlation between human activities and soil salinity dates back centuries, with documented instances of civilizations collapsing due to heightened salinity levels, as evidenced by the downfall of Mesopotamia. Whether through natural processes or human mismanagement, soil salinity degrades soil quality, potentially compromising its self-regulating capacity.

The prevalence of soil salinity is widespread, impacting more than 100 countries globally, with no continent being spared. Future projections indicate a worsening of salinization due to climate change-related factors such as rising sea levels and increased temperatures, leading to escalated evaporation rates and subsequent salinization. This phenomenon can disrupt ecosystems, impeding their ability to fulfill crucial environmental functions. While recent global data on soil salinization are scarce, it is reasonable to infer an expansion of the issue since data collection in the 1970s and 1980s, surpassing reclamation efforts. Numerous countries grapple with salt-induced land degradation, including regions like the Aral Sea Basin in Central Asia, the Indo-Gangetic Basin in India, and the Murray-Darling Basin in Australia, among others.

Conclusion

The historical perspective on problem soils reveals a long-standing struggle to understand and manage these challenging agricultural and environmental resources. From ancient civilizations employing primitive reclamation techniques to modern scientific advancements in soil chemistry and microbiology, the journey has been marked by continuous learning and adaptation. Early farmers recognized the limitations of saline, acidic, or waterlogged soils, while later agricultural revolutions introduced lime, gypsum, and organic amendments to mitigate soil degradation. In the 20th and 21st centuries, the focus shifted toward sustainable practices, including phytoremediation, precision agriculture, and climate-resilient soil management. Historical experiences underscore the importance of interdisciplinary approaches—combining traditional knowledge with modern technology—to restore soil health and ensure food security. As population pressures and climate change exacerbate soil problems, lessons from history emphasize the need for proactive, science-based solutions to preserve this vital natural resource for future generations. Thus, understanding the historical evolution of problem soil management provides valuable insights for developing innovative strategies to combat soil degradation and enhance global agricultural productivity sustainably.

References

Abrol, I. P., Saha, A. K., & Acharya, C. L. (1978). Effect of exchangeable sodium on some soil physical properties. Journal of the Indian Society of Soil Science, 26(2), 98-105.

Acharya, C. L., & Abrol, I. P. (1976). Effect of river sand on the permeability of a sodic soil. J. Indian Soc. Soil Sci, 24, 245-252.

Acharya, C. L., & Abrol, I. P. (1991). Soil water relation and root water extraction in alkali soils.

Baldwin, M., Kellogg, C. E., & Thorp, J. (1938). Soil classification (pp. 979-1001). Indianapolis, Indiana: Bobbs-Merrill.

Brevik, E. C., & Hartemink, A. E. (2010). Early soil knowledge and the birth and development of soil science. Catena, 83(1), 23-33.

Brevik, E. C., Homburg, J. A., Miller, B. A., Fenton, T. E., Doolittle, J. A., & Indorante, S. J. (2016). Selected highlights in American soil science history from the 1980s to the mid-2010s. Catena, 146, 128-146.

Das, B. M. (2019). Advanced soil mechanics. CRC press.

Das, D. K. (2011). Introduction to soil science. Kalyani Publication, India, 645.

Fitzpatrick, R. W. (2013). Demands on soil classification and soil survey strategies: special-purpose soil classification systems for local practical use. Developments in soil classification, land use planning and policy implications: innovative thinking of soil inventory for land use planning and management of land resources, 51-83.

Jameson, W. (1853). Contributions to a history of the relation between climate and vegetation on various parts of the globe: on the physical aspect of the Punjab its agriculture and Botany. The Journal of the Horticultural Society of London, 8, 273-314.

Lal, R. (2017). Encyclopedia of soil science. CRC Press.

Moreland, W. H., & Tamhane, U. A. (1901). Reh. Agric Ledger, 13, 416-463.

Rao, D. L. N., Singh, N. T., Gupta, R. K., & Tyagi, N. K. (Eds.). (1994). Salinity management for sustainable agriculture (25 years of research at CSSRI).

Russell, R. S., & Goss, M. J. (1974). Physical aspects of soil fertility-the response of roots to mechanical impedance. Netherlands Journal of Agricultural Science, 22(4), 305-318.

Sharma, R. C., Mandal, A. K., Singh, R., & Singh, Y. P. (2011). Characteristics and use potential of sodic and associated soils in CSSRI experimental farm, Lucknow, Uttar Pradesh. Journal of the Indian Society of Soil Science, 59(4), 381-387.

Tan, K. H. (2009). Environmental soil science. CRC Press.

Tiwari, S. C., Tiwari, B. K., & Mishra, R. R. (1989). Microbial populations, enzyme activities and nitrogen-phosphorus-potassium enrichment in earthworm casts and in the surrounding soil of a pineapple plantation. Biology and Fertility of Soils, 8, 178-182.

van den Ban, A. W. (1994). Possible roles of the extension group of the technology evaluation and transfer division at CSSRI: a discussion paper for the Indo-Dutch soil salinity and reclamation project. In Extension education: a reader for BIOCON (No. July 1994, pp. 1-12). Wageningen Agricultural University.

2

Soil Quality and Soil Health

Introduction

Soil quality is a critical concept in environmental science, agriculture, and land management. It refers to the ability of soil to perform its essential functions, which include supporting plant growth, regulating water flow, cycling nutrients, filtering pollutants, and providing habitat for microorganisms. Healthy, high-quality soil is fundamental for the productivity of ecosystems and the sustainability of agricultural practices. The quality of soil is influenced by a combination of physical, chemical, and biological factors. These include soil texture, structure, organic matter content, pH, nutrient availability, and the presence of beneficial organisms like bacteria and fungi. When these factors are balanced, soil can support healthy crops and ecosystems. However, soil degradation—caused by practices such as overuse of chemical fertilizers, deforestation, erosion, and pollution—can severely reduce its quality, leading to lower agricultural yields, water contamination, and loss of biodiversity. Assessing soil quality involves monitoring several key indicators, such as organic matter levels, soil texture, compaction, water infiltration rates, and the presence of key nutrients. Organic matter is especially important, as it helps to bind soil particles, retain moisture, and provide a food source for beneficial microorganisms. The chemical aspect of soil quality is equally vital, encompassing the availability of nutrients such as nitrogen, phosphorus, and potassium, as well as the presence of any harmful contaminants like heavy metals. The biological dimension of soil quality relates to the diversity and activity of organisms within the soil. A rich microbial community enhances nutrient cycling, helps decompose organic matter, and promotes plant health. Earthworms, for instance, play a critical role in aerating the soil and facilitating water infiltration. Soil quality is not a static property; it can change over time depending on land management practices. Sustainable practices such as crop rotation, conservation tillage, the use of organic amendments, and maintaining ground cover can improve and maintain soil quality. Conversely, unsustainable practices like excessive tillage, monoculture, and the overuse of agrochemicals can degrade soil, making it less productive and more prone to erosion. In recent years, the importance of soil quality has gained significant attention due to its direct link to food security and environmental health. The growing global population and the increasing demand for agricultural productivity put pressure on soil resources. As such, maintaining soil quality is essential for ensuring long-term agricultural sustainability, protecting water resources, and mitigating the effects of climate change. Sustainable soil management practices can enhance the resilience of ecosystems, contribute to carbon sequestration, and reduce greenhouse gas emissions.

The term and concept of soil quality evokes various responses depending on our scientific and social backgrounds. For some, soil quality evokes an ethical or emotional tie to the land. To others, soil quality is an integration of soil processes and provides a measure of change in soil condition as related to factors such as land use, climate patterns, cropping sequences and farming systems. Most comprehensive and accepted definition of soil quality is the one given by **Soil Science Society of America (Karlen *et al.* 1997)** which inter alia reads "*Soil quality is the capacity of a specific kind of soil to function within natural or managed ecosystem boundaries to sustain plant and animal productivity, maintain or enhance water and air quality and support human health and habitation*".

Soil Quality = inherent soil quality (Soil suitability) + Dynamic soil quality (Soil health)

Factors effecting soil quality

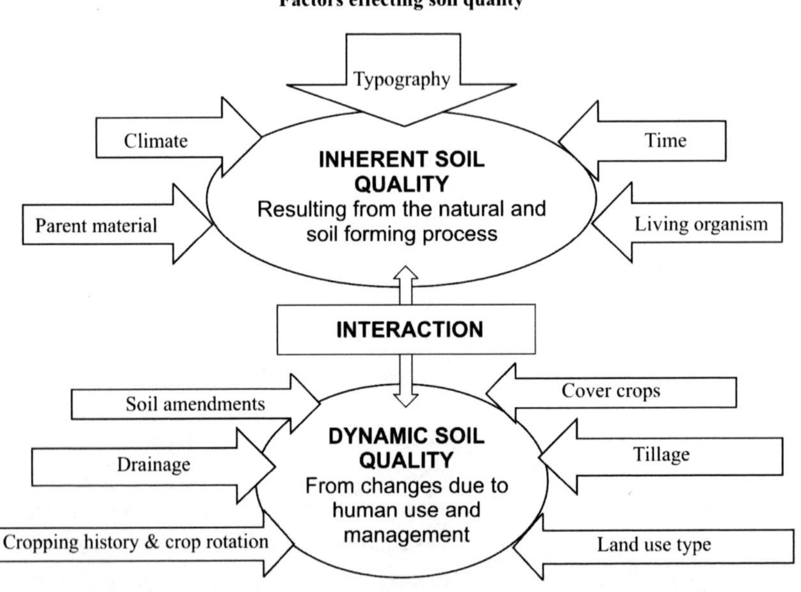

Source: Soil Science Society of America Journal (1997)

Inherent soil quality (soil suitability)

Inherent soil quality refers to the natural properties of the soil that are determined by long-term factors such as its parent material, climate, topography, and time of formation. These qualities are largely stable over time and define the soil's baseline capacity to support plant growth, water regulation, and nutrient cycling. Inherent soil properties include soil texture, depth, natural fertility, and drainage capacity, which are not easily altered through management practices.

Characteristics of Inherent Soil Quality

Texture: Soils have a natural texture (the proportion of sand, silt, and clay) that determines their water retention and drainage capabilities.

Depth: The natural depth of soil to bedrock or a restrictive layer influences root penetration and water storage capacity.

Mineral Composition: The natural availability of essential nutrients (such as phosphorus or potassium) is largely determined by the minerals present in the parent material.

Natural pH: The soil's baseline pH, influenced by its mineral content and the local climate, affects nutrient availability and microbial activity.

Drainage Capacity: Some soils, such as sandy soils, naturally drain quickly, while clayey soils retain water for longer periods.

Dynamic soil quality

Dynamic soil quality refers to the properties of soil that can change over time due to human activities or environmental factors. It includes soil characteristics such as organic matter content, nutrient availability, soil structure, and biological activity. Unlike inherent qualities, dynamic soil quality can be improved or degraded relatively quickly through land management practices, such as tillage, crop rotation, or organic amendments. Dynamic soil quality reflects the soil's current health and capacity to perform its functions effectively.

Characteristics of Dynamic Soil Quality

Organic Matter Content: Soil organic matter is a key component of dynamic soil quality. It can be increased through the addition of compost, cover crops, and reduced tillage practices.

Nutrient Availability: Soil fertility (the availability of essential nutrients) can be enhanced or diminished through fertilizer application, crop rotation, or depletion from continuous monoculture.

Soil pH: Soil acidity or alkalinity can change over time, especially with the addition of lime (to reduce acidity) or sulfur (to increase acidity). Dynamic soil quality management aims to maintain optimal pH for plant growth.

Biological Activity: The activity of soil organisms (bacteria, fungi, earthworms) is a dynamic indicator of soil health. Management practices such as organic amendments, cover cropping, and minimizing pesticide use can increase biological activity.

Comparison of Inherent and Dynamic Soil Quality

Aspect	Inherent Soil Quality	Dynamic Soil Quality
Definition	Natural, baseline properties of soil that are largely determined by formation factors (parent material, climate, topography, time).	Properties of soil that can change based on land management and environmental conditions.
Key Factors	Parent material, climate, topography, soil age.	Land management practices, organic matter content, compaction, nutrient management.
Stability	Relatively stable over time; not easily altered by human intervention.	Can change rapidly in response to human activities and environmental conditions.
Examples	Soil texture, natural depth to bedrock, inherent mineral composition.	Organic matter content, soil structure, nutrient availability, biological activity.
Improvement Potential	Limited ability to improve inherent qualities without major interventions.	Can be improved or degraded through proper management, such as organic amendments or reduced tillage.
Impact on Productivity	Sets the natural potential for productivity and ecosystem services.	Directly affects the current health and productivity of the soil.

Concepts of Soil Quality

Soil quality refers to the capacity of soil to function effectively within natural or managed ecosystems, supporting biological productivity, maintaining environmental quality, and promoting plant and animal health. Unlike soil fertility, which focuses predominantly on the nutrient status for plant growth, soil quality encompasses a holistic view of the soil's ability to perform multiple functions simultaneously. It is the foundation for sustainable agriculture, environmental conservation, and ecosystem resilience.

Key functions of soil include

Supporting Plant and Animal Productivity: Healthy soils provide a medium for plants to grow by supplying essential nutrients, water, and support for roots. These soils are vital for food production, forestry, and biodiversity.

Regulating Water: Soil acts as a filter and storage medium for water, regulating its flow and availability to plants and animals. High-quality soils enhance water infiltration, retention, and filtration.

Nutrient Cycling: Soil plays a crucial role in cycling essential nutrients such as nitrogen, phosphorus, and carbon, supporting both plant growth and ecosystem sustainability.

Filtering and Buffering Pollutants: Soil filters pollutants and acts as a buffer, reducing the impact of harmful chemicals and environmental toxins. Well-functioning soils prevent contamination of water bodies by immobilizing or degrading pollutants.

Supporting Biodiversity and Habitat: Soil provides a habitat for a diverse range of microorganisms, insects, and animals, which play key roles in maintaining soil structure, nutrient cycling, and ecosystem services.

Principles of Soil Quality

Several key principles guide the understanding and management of soil quality. These principles emphasize the need to maintain or improve soil's biological, chemical, and physical integrity to ensure long-term sustainability:

Soil as a Living System: Healthy soil is a biologically active system, home to a vast array of organisms, including bacteria, fungi, earthworms, and insects. These organisms contribute to nutrient cycling, organic matter decomposition, and soil structure. Enhancing biological diversity and activity is fundamental for maintaining soil quality.

Balance between Soil Components: The quality of soil is determined by the balance among its physical, chemical, and biological components. Disruptions in one aspect can affect the others. For example, excessive use of fertilizers may alter the chemical balance, reducing microbial diversity and soil structure.

Soil Resilience: High-quality soils have a greater ability to resist and recover from stresses, such as erosion, compaction, drought, and pollution. The resilience of soil depends on its organic matter content, structure, and biological activity, which enable the soil to adapt to changes in environmental conditions.

Soil Degradation Prevention: Once soil is degraded, restoring its quality can be a slow and difficult process. Therefore, soil conservation practices should focus on preventing degradation, such as erosion, nutrient depletion, and compaction, rather than attempting to reverse damage after it occurs.

Sustainability: Soil quality management should be aimed at sustainable land use, ensuring that soil remains productive for future generations. This includes adopting agricultural and land management practices that maintain or improve soil's capacity to function over time without degrading its essential qualities.

Components of Soil Quality

Soil quality is determined by the interaction of several components, broadly categorized into physical, chemical, and biological properties. Each component is integral to how well the soil performs its multiple functions.

1. Physical Components

The physical characteristics of soil are critical for supporting plant roots, regulating water flow, and maintaining soil structure.

Soil Texture: Soil texture refers to the relative proportions of sand, silt, and clay particles in the soil. It affects water retention, drainage, and root penetration. Sandy soils have good drainage but low water retention, while clayey soils retain more water but may have poor drainage and aeration.

Soil Structure: Soil structure describes how soil particles aggregate or clump together, forming pores for air and water movement. Well-aggregated soils with stable structure promote good water infiltration, root growth, and resistance to erosion. Poor structure, often caused by compaction, restricts root development and limits plant access to water and nutrients.

Bulk Density: Bulk density measures the mass of soil in a given volume and is an indicator of soil compaction. High bulk density, often caused by heavy machinery or excessive grazing, reduces pore space, impeding water infiltration, root growth, and soil aeration. Low bulk density is associated with high porosity and better soil structure.

Water-Holding Capacity: The ability of soil to retain water is essential for plant growth, particularly in dry conditions. Soil with high organic matter and good structure holds water more effectively, reducing the need for irrigation and protecting plants during drought.

Porosity: The presence of pore spaces in the soil allows air and water to move freely, promoting root growth and microbial activity. High porosity is associated with better soil aeration and water infiltration, while low porosity can lead to waterlogging and poor root health.

2. Chemical Components

The chemical properties of soil determine its fertility, nutrient availability, and pH balance, all of which are crucial for plant growth and ecosystem function.

Soil pH: Soil pH measures the acidity or alkalinity of the soil, which influences nutrient availability and microbial activity. Most plants grow best in soils with a pH between 6 and 7.5, as extreme pH levels can lead to nutrient deficiencies or toxicities.

Nutrient Availability: Essential nutrients such as nitrogen (N), phosphorus (P), potassium (K), and micronutrients (e.g., zinc, iron) must be present in adequate amounts and available to plants. Nutrient imbalances or deficiencies can limit plant growth, while excess nutrients, particularly nitrogen and phosphorus, can lead to environmental pollution through runoff and leaching.

Cation Exchange Capacity (CEC): CEC refers to the soil's ability to retain and exchange positively charged ions (cations) such as calcium, magnesium, potassium, and ammonium. Soils with higher CEC can hold more nutrients, making them more fertile. Soils with low CEC, such as sandy soils, are less able to retain nutrients, leading to leaching and nutrient loss.

Salinity and alkalinity of soil: High levels of soluble salts in the soil, often a result of poor irrigation management or saline water sources, can affect plant growth by disrupting water uptake and nutrient availability. Salinity is a major issue in arid and semi-arid regions.

3. Biological Components

The biological properties of soil are driven by the activity of microorganisms, fungi, and fauna, which contribute to nutrient cycling, organic matter decomposition, and soil structure.

Soil Organic Matter (SOM): SOM includes decomposed plant and animal residues, living and dead microorganisms, and other organic substances. It is a key indicator of soil health and fertility, providing nutrients, enhancing water retention, and improving soil structure. High levels of organic matter also promote carbon sequestration, which helps mitigate climate change.

Microbial Activity: Soil is home to a diverse community of microorganisms, including bacteria, fungi, and protozoa, which play essential roles in nutrient cycling, organic matter decomposition, and disease suppression. Active microbial communities contribute to soil fertility and plant health by breaking down organic matter, releasing nutrients, and forming symbiotic relationships with plant roots (e.g., *mycorrhizae*).

Earthworms and Soil Fauna: Earthworms and other soil fauna, such as insects and nematodes, are important for maintaining soil structure and promoting nutrient cycling. Earthworms improve soil aeration, water infiltration, and organic matter breakdown through their burrowing activities. Their presence is often an indicator of healthy soil.

Root Systems: Plant roots contribute to soil structure by creating channels that improve water and air movement. Roots also exude organic compounds that feed soil microorganisms and promote nutrient cycling.

Assessment and Evaluation of Soil Quality

Assessing and evaluating soil quality involves determining the capacity of soil to perform its functions, which include supporting plant growth, regulating water flow, cycling nutrients, filtering pollutants, and sustaining biodiversity. The process combines scientific methods to measure physical, chemical, and biological soil properties, with an interpretation of those results based on soil function.

Assessment vs. Evaluation

Soil Quality Assessment: Refers to the process of measuring specific soil properties (indicators) to provide data on the soil's health. It is focused on quantifying soil conditions using various tests and sampling methods.

Soil Quality Evaluation: Involves interpreting the data gathered from the assessment process. Evaluation is more qualitative, determining whether the soil is suitable for specific land uses, such as agriculture, forestry, or environmental protection.

Methods of Soil Quality Assessment

Soil quality assessment requires selecting appropriate indicators, sampling the soil, and using either field or laboratory tests to quantify the data. Below is a detailed overview of common methods used to assess soil quality.

1. Physical Methods

Physical indicators assess the soil's structural integrity, its ability to retain water, and its resilience to erosion and compaction. Common physical properties measured in soil quality assessments include:

Soil Texture

Method: The soil texture triangle method (field test) or particle size analysis (laboratory).

Purpose: Determines the proportion of sand, silt, and clay, which influences water retention, drainage, and aeration.

Bulk Density

Method: Soil core sampling followed by weighing the dried sample.

Purpose: High bulk density indicates compaction, which reduces porosity, water infiltration, and root growth.

Water Holding Capacity

Method: Gravimetric water content test, using soil samples and calculating the moisture retention at field capacity.

Purpose: Measures the soil's ability to store water, essential for plant growth during dry periods.

Soil Structure and Aggregate Stability

Method: Visual assessment of soil aggregates or the wet-sieving method in a laboratory.

Purpose: Evaluates how well soil particles are held together. Good structure supports water infiltration and root development.

Soil Erosion Risk

Method: Using erosion pins in the field or remote sensing technologies.

Purpose: Evaluates the soil's susceptibility to wind and water erosion. It also assesses soil cover and slope.

2. Chemical Methods

Chemical properties are crucial for understanding the nutrient content and potential fertility of soil. Common chemical indicators include:

Soil pH

Method: Field pH meters or laboratory titration.

Purpose: pH controls nutrient availability and microbial activity. Ideal pH ranges for most crops are between 6.0 and 7.5.

Cation Exchange Capacity (CEC)

Method: Laboratory analysis using ammonium acetate or barium chloride to measure the soil's ability to hold and exchange cations.

Purpose: CEC indicates soil fertility and nutrient-holding capacity. High CEC soils can retain more nutrients, whereas low CEC soils may need frequent fertilization.

Nutrient Levels (N, P, K)

Method: Soil extraction methods such as Olsen-P for phosphorus or Kjeldahl analysis for nitrogen.

Purpose: Measures the availability of essential plant nutrients. Deficiencies or excesses of nutrients can limit crop growth or cause environmental harm.

Soil Organic Matter (SOM)

Method: Loss-on-ignition (LOI) or Walkley-Black method.

Purpose: SOM influences nutrient cycling, water retention, and soil structure. Increasing SOM generally improves soil health.

Salinity

Method: Electrical conductivity (EC) measurement.

Purpose: High soil salinity can reduce crop yield and soil health, especially in arid and semi-arid regions.

3. Biological Methods

Biological indicators of soil quality assess the abundance and activity of soil organisms. Soil biology is vital for nutrient cycling, organic matter decomposition, and maintaining soil structure.

Soil Respiration

Method: Incubation of soil samples and measurement of CO_2 release using gas analysis techniques.

Purpose: Measures microbial activity, which is a key indicator of soil biological health. Higher respiration rates often indicate healthy microbial populations.

Soil Microbial Biomass

Method: Chloroform fumigation extraction method.

Purpose: Assesses the quantity of living microbial organisms in the soil, which reflects soil fertility and its ability to cycle nutrients.

Earthworm Population

Method: Counting earthworms through digging and sieving or by using earthworm traps.

Purpose: Earthworms improve soil structure, aeration, and nutrient cycling. A healthy soil typically has abundant earthworms.

Soil Enzymatic Activity

Method: Laboratory analysis to measure enzyme activity (e.g., dehydrogenase, phosphatase).

Purpose: Enzymatic activity reflects microbial processes related to organic matter decomposition and nutrient cycling.

4. Visual Soil Assessment (VSA)

VSA is a rapid, cost-effective field method for assessing soil quality. It involves visual observations of key indicators such as soil color, structure, compaction, and root development.

Method

Observe soil tilth, color, presence of plant residues, rooting depth, and compaction layers.

Assess soil quality through simple tests such as a ribbon test for texture or a spade test for structure.

Purpose: Provides an immediate, qualitative understanding of soil health without the need for laboratory analysis.

Soil Quality Evaluation: Methods

After gathering the data from the assessment phase, soil quality is evaluated by comparing the results against benchmarks, reference soils, or desired functions. The following are methods used for evaluation:

1. Soil Quality Indexes (SQI)

Soil quality indexes combine multiple indicators into a single score that reflects the overall soil quality for a given purpose (e.g., crop productivity or environmental health).

Method: Select and weight key indicators based on their importance to the soil's functions.

Normalize data (convert indicators to a common scale)

Use a formula or model (often additive or weighted) to calculate a composite index score.

Purpose: SQIs simplify the interpretation of soil quality data and make it easier to communicate results to stakeholders such as farmers, policymakers, or land managers.

2. Benchmarking and Trend Analysis

Method: Compare current soil quality measurements with baseline data, local standards, or historical trends. Identify whether the soil quality is improving, degrading, or stable over time.

Purpose: Benchmarking provides a reference for whether soil management practices are sustainable or need to be adjusted to improve soil health.

3. Land Suitability and Capability Classification

Land capability classification evaluates soil quality in terms of its potential for specific land uses, such as agriculture or construction.

Method: Classify soil into different capability classes based on its limitations (e.g., slope, drainage, fertility).

Use tools such as the USDA land capability classification system.

Purpose: Helps in land-use planning and in selecting appropriate management practices based on soil limitations and potential.

Strategies for improving soil quality

The properties of soil which represent the dynamic soil quality can be improved by several management practices which are described as follows:

Enhancement of organic matter: Organic matter is considered as the main stay of good soil quality. Regular additions of organic matter improve soil structure, enhance water and nutrient holding capacity, protect soil from erosion, hard setting and compaction and support a healthy community of soil organisms.

Reduction in the intensity of tillage: Reducing tillage minimizes the loss of organic matter and protects the soil surface with plant residue. Tillage is used to loosen surface soil, prepare the seedbed and control weeds and pests. But tillage can also break up soil structure, speed up the decomposition and loss of organic matter, increase the threat of erosion, destroy the habitat of helpful organisms and cause compaction.

Efficient management of pests and nutrients: Efficient pest and nutrient management means testing and monitoring soil and pests; applying only the necessary chemicals, at the right time and place to get the job done; and taking advantage of non-chemical approaches to pest and nutrient management such as crop rotations, cover crops and manure management. The terms integrated pest management (IPM) and integrated nutrient managements (INM) are very much popular nowadays.

Prevention of soil compaction: Soil compaction reduces the amount of air, water and space available to roots and soil organisms. Compaction is caused by repeated traffic, heavy traffic or traveling on wet soil.

Maintenance of ground cover: Soil without adequate cover or bare soil is very much susceptible to wind and water erosion and to drying and crusting. Ground cover protects soil; provides habitats for larger soil organisms, such as insects and earthworms and can improve water availability.

Diversification of cropping systems: Diversity is beneficial for several reasons. Each plant contributes a unique root structure and type of residue to the soil. A diversity of soil organisms can help control pest populations and a diversity of cultural practices can reduce weed and disease pressures.

Factors Affecting Soil Quality

Soil quality, defined as the capacity of soil to function effectively within an ecosystem to support plant growth, regulate water, cycle nutrients, and sustain biodiversity, is shaped by a combination of natural and human-induced factors. These factors affect the soil's physical, chemical, and biological properties and, consequently, its ability to perform essential functions. Understanding these factors is critical for managing soil health and ensuring sustainable land use. This chapter discusses the natural and human-induced factors affecting soil quality in detail.

1. Natural Factors Affecting Soil Quality

Natural factors are those that primarily define the inherent soil quality, which is relatively stable over time and difficult to change without significant alteration to the environment. These factors are largely determined by natural processes such as soil formation, weathering, and vegetation growth.

a) Parent Material

The parent material is the geological or organic material from which the soil is formed. It is the foundational layer that influences the mineral composition, texture, and fertility of the soil.

Impact on Soil Quality: Soils formed from coarse parent materials like granite or sandstone are typically more acidic and less fertile, while those derived from fine-textured materials like basalt or limestone tend to be more nutrient-rich and fertile. Parent material determines the soil's mineral content, which in turn influences its nutrient availability and fertility. For example, soils derived from limestone are often high in calcium, contributing to higher pH and better nutrient retention.

b) Climate

Climate, particularly temperature and precipitation, plays a central role in soil formation, weathering rates, and biological activity within the soil.

Impact on Soil Quality

Temperature: Warmer climates accelerate organic matter decomposition and enhance microbial activity, promoting nutrient cycling and soil fertility. Conversely, cooler climates slow down these processes, leading to higher organic matter accumulation but reduced nutrient availability.

Precipitation: High rainfall can lead to leaching of nutrients like nitrogen and potassium, making soils more acidic and less fertile. In arid regions, insufficient rainfall can lead to salinization, where salts accumulate in the soil, degrading its quality and reducing plant growth. Seasonal variations in temperature and rainfall also impact erosion rates, microbial activity, and the soil's water-holding capacity.

c) Topography

Topography refers to the shape and slope of the land, which affects water movement, soil erosion potential, and soil depth.

Impact on Soil Quality: Soils on steep slopes are more susceptible to erosion, leading to the loss of the nutrient-rich topsoil layer. This reduces soil fertility and structure. Valleys and flat areas often accumulate organic matter, minerals, and nutrients from surrounding higher areas, which improves soil fertility. However, these areas may also experience poor drainage or waterlogging, depending on the local climate.

The slope aspect (north-facing vs. south-facing slopes) influences solar radiation, affecting soil temperature and moisture levels, which in turn influence plant growth and microbial activity.

d) Vegetation

Vegetation contributes organic matter to the soil, influences water retention, and promotes biological activity through root systems and microbial interactions.

Impact on Soil Quality: Soils under forest cover tend to have a higher organic matter content and better structure due to the continuous input of leaf litter and root biomass. Forest soils often support diverse microbial communities that enhance nutrient cycling and soil health. Grassland soils are often rich in organic matter due to the dense root systems of grasses. These soils typically have good aggregation, water infiltration, and nutrient cycling. Different plant species also influence the soil's chemical properties. For instance, nitrogen-fixing legumes increase soil nitrogen content, enhancing fertility.

e) Time

The age of soil, or the amount of time it has been exposed to weathering and biological activity, affects its development and quality.

Impact on Soil Quality: Older soils tend to be more weathered and leached of essential nutrients, especially in tropical regions with high rainfall. Such soils may require replenishment of nutrients for sustained productivity. Younger soils, such as those recently formed from volcanic activity or glacial deposits, may be less weathered and rich in nutrients, making them initially more fertile.

2. Human-Induced Factors Affecting Soil Quality

Human activities significantly influence dynamic soil quality, which refers to soil properties that can change over time due to management practices. These factors have the potential to either improve or degrade soil health, depending on how land is used and managed.

a) Land Use and Management Practices

Land use refers to how the land is utilized (e.g., agriculture, forestry, urban development), and management practices include the specific techniques used to manage soils (e.g., tillage, crop rotation).

Impact on Soil Quality

Tillage: Excessive or inappropriate tillage can break down soil structure, leading to compaction, erosion, and reduced organic matter. Conservation tillage or no-till practices help maintain soil structure, increase organic matter retention, and reduce erosion.

Monoculture: Growing the same crop repeatedly on the same land depletes specific nutrients, reduces biodiversity, and increases vulnerability to pests and diseases. Crop rotation and diversified cropping systems improve nutrient cycling, enhance biodiversity, and build organic matter in the soil.

Deforestation: The removal of trees for agriculture or urban development reduces organic matter input, increases erosion, and can lead to compaction from heavy machinery.

Urbanization: Expanding urban areas result in soil sealing (paving over soil), which prevents water infiltration, disrupts natural soil processes, and diminishes soil's ability to function effectively.

b) Organic Matter and Residue Management

Organic matter management involves the addition and maintenance of organic materials, such as crop residues, compost, and manure, to enhance soil properties.

Impact on Soil Quality: Organic matter improves soil structure, increases water retention, enhances microbial activity, and promotes nutrient availability. Regular application of organic inputs like compost or manure helps build and maintain soil fertility over time. Removing crop residues (e.g., through burning) depletes the soil of valuable organic matter, reduces water retention, and increases erosion risk.

c) Irrigation Practices

Irrigation involves the artificial application of water to soil to support plant growth, particularly in arid or semi-arid regions.

Impact on Soil Quality

Over-irrigation: Can lead to waterlogging, reducing soil aeration and impairing root growth. Prolonged over-irrigation may also cause soil salinization, where salts accumulate in the root zone, degrading soil fertility and plant productivity.

Efficient irrigation: Practices like drip irrigation or sprinklers minimize water waste, improve water-use efficiency, and reduce the risk of salinization. This also maintains optimal soil moisture levels for healthy plant growth.

d) Chemical Inputs (Fertilizers, Pesticides, Herbicides)

The use of chemical inputs, such as fertilizers, pesticides, and herbicides, is common in modern agriculture to increase crop yields and manage pests.

Impact on Soil Quality

Fertilizers: While fertilizers replenish essential nutrients, overuse can result in nutrient imbalances (e.g., excess nitrogen), leading to nutrient leaching, pollution, and soil acidification. Excessive fertilizer use can also disrupt the soil's natural nutrient cycling processes.

Pesticides and Herbicides: Heavy use of chemical pesticides can harm beneficial soil organisms, reduce biodiversity, and disrupt the natural balance of soil ecosystems. Pesticides may also contaminate water sources and lead to the buildup of toxic residues in the soil.

e) Soil Compaction

Soil compaction occurs when soil particles are pressed together, reducing pore space, restricting root growth, and limiting water and air movement in the soil.

Impact on Soil Quality

Heavy machinery: Repeated use of heavy machinery in agriculture, construction, or forestry can lead to soil compaction, especially in wet conditions. Compacted soils have reduced porosity, limiting root development and microbial activity.

Livestock overgrazing: Overgrazing by livestock can compact soil, especially around water points or feeding areas, leading to reduced infiltration, increased runoff, and higher erosion risks.

f) Soil Erosion

Erosion refers to the removal of the topsoil layer by wind or water, leading to the loss of fertile soil.

Impact on Soil Quality

Water erosion: Results from heavy rainfall or improper irrigation practices, causing the loss of topsoil and nutrients. This depletes soil fertility and weakens soil structure, making it less productive.

Wind erosion: Occurs in dry, sandy soils and areas without vegetation cover, stripping away fine particles and organic matter that are essential for soil fertility.

Conservation practices: Implementing soil conservation techniques, such as contour farming, cover cropping, and terracing, can mitigate erosion risks and preserve soil quality.

g) Pollution and Contamination

Soil contamination occurs when harmful substances, such as heavy metals, industrial pollutants, or agricultural chemicals, accumulate in the soil.

Impact on Soil Quality

Heavy metals: Accumulation of heavy metals from industrial pollution, mining, or improper waste disposal can poison plants and soil organisms, reducing soil fertility and biological activity.

Soil Health

Soil health is a fundamental aspect of environmental science and agriculture, referring to the ability of soil to function as a living, dynamic ecosystem that sustains plants, animals, and humans. Unlike traditional measures of soil fertility, which focus solely on nutrient levels, soil health encompasses a broader range of biological, physical, and chemical properties that work together to support life and maintain ecosystem services. Healthy soil is resilient, productive, and able to recover from stress while providing essential ecosystem functions such as nutrient cycling, water filtration, and carbon sequestration.

Key Components of Soil Health

Soil health is determined by several interrelated factors:

Physical Properties: The physical structure of soil plays a significant role in its health. Soil texture, which refers to the proportion of sand, silt, and clay particles, influences water retention, drainage and root penetration. Soil aggregation, or the way soil particles clump together, affects porosity and aeration, which are critical for root growth and microbial activity. Compacted soils, with reduced pore spaces, limit water infiltration and oxygen availability, both of which are essential for healthy plant growth and microbial processes.

Chemical Properties: Chemical aspects of soil health primarily involve nutrient availability and balance. Essential nutrients like nitrogen (N), phosphorus (P), and potassium (K) are vital for plant growth, and their presence in the right proportions is crucial for soil health. Soil pH, which influences nutrient availability, is also a key factor. A healthy soil should have balanced levels of nutrients and a pH range suitable for plant growth, typically between 6 and 7 for most crops. However, the overuse of chemical fertilizers, improper pH levels, or the presence of toxic elements like heavy metals can severely impact soil health, reducing its productivity and ability to support ecosystems.

Biological Properties: Soil is teeming with life, from microorganisms such as bacteria and fungi to larger organisms like earthworms and insects. This biological activity is a key indicator of soil health. Microorganisms play a crucial role in decomposing organic matter, cycling nutrients, and enhancing soil structure. Earthworms, for instance, help aerate the soil and improve its physical properties, while nitrogen-fixing bacteria convert atmospheric nitrogen into forms plants can use. A diverse and active soil biological community is essential for nutrient cycling, organic matter decomposition, and the suppression of pests and diseases. The loss of biological diversity or activity in the soil can result in poor nutrient cycling and diminished resilience.

Organic Matter: Soil organic matter is a cornerstone of soil health, contributing to both its physical structure and its capacity to support biological activity. It consists of decomposed plant and animal materials and provides a vital food source for microorganisms. Organic matter improves soil structure by binding soil particles together, enhancing water retention, and increasing nutrient availability. It also plays a role in carbon sequestration, helping mitigate climate change by storing carbon that would otherwise be released into the atmosphere as CO_2. Maintaining or increasing organic matter levels through practices such as composting or cover cropping is critical for long-term soil health.

Indicators of Soil Health

To assess soil health, scientists and farmers rely on several indicators:

Soil Organic Matter (SOM): A higher percentage of organic matter indicates a well-functioning soil ecosystem, with good nutrient cycling, water retention, and biological activity.

Soil Respiration: This measures the activity of microorganisms by assessing the amount of CO they release during organic matter decomposition. Higher respiration rates generally indicate a more active and healthy microbial community.

Aggregate Stability: The ability of soil to form and maintain aggregates is an indicator of good soil structure, which is essential for water infiltration, root growth, and erosion prevention.

Nutrient Availability: Testing for key nutrients such as nitrogen, phosphorus, and potassium helps determine whether the soil can support plant growth. A healthy balance of nutrients is essential.

pH Levels: Soil pH affects the availability of nutrients. Maintaining pH in the optimal range for specific crops ensures that nutrients remain accessible to plants.

Factors Affecting Soil Health

Several factors can either promote or degrade soil health:

Land Management Practices: Sustainable land management is crucial for maintaining soil health. Practices such as crop rotation, conservation tillage, and the use of cover crops can protect soil structure, enhance nutrient cycling, and promote biodiversity. Conversely, practices like monoculture, overgrazing, excessive tillage, and the overuse of agrochemicals can deplete soil nutrients, reduce organic matter, and disrupt biological activity.

Erosion and Compaction: Soil erosion caused by wind or water removes the topsoil layer, which is the most fertile part of the soil. This loss of topsoil can severely reduce soil health by depleting nutrients and organic matter. Soil compaction, often caused by heavy machinery or overgrazing, reduces pore spaces in the soil, hindering water infiltration and root penetration.

Chemical Inputs: The overuse of chemical fertilizers and pesticides can have detrimental effects on soil health. While these inputs may increase short-term crop yields, they can disrupt the balance of nutrients in the soil, reduce organic matter, and negatively affect microbial communities. Excessive chemical inputs may also lead to soil acidification, salinization, and contamination with toxic elements like heavy metals.

Climate Change: Changes in climate, such as increased temperatures, altered precipitation patterns, and more frequent extreme weather events, can impact soil health. For instance, more intense rainfall may increase erosion, while prolonged droughts can reduce microbial activity and organic matter decomposition.

Importance of Soil Health

Healthy soil is the foundation of sustainable agriculture and ecosystem resilience. It provides a medium for plant growth, regulates water and air flow, filters pollutants, and supports diverse biological communities. Healthy soils also play a significant role in mitigating climate change by storing carbon in the form of organic matter and reducing greenhouse gas emissions. In agriculture, soil health is directly linked to crop productivity and food security. Healthy soils enhance plant growth by providing essential nutrients, improving water retention, and reducing the incidence of pests and diseases. Maintaining soil health is essential for long-term agricultural sustainability, particularly as the global population continues to grow and demand for food increases.

Moreover, healthy soils contribute to environmental sustainability by filtering pollutants, regulating water quality, and preventing erosion. By maintaining healthy soils, we can help protect natural ecosystems, preserve biodiversity, and ensure clean water supplies.

Difference between Soil Quality and Soil Health

Aspect	Soil Quality	Soil Health
Definition	The capacity of soil to function and perform specific roles in an ecosystem, such as supporting plant growth, water regulation, and nutrient cycling.	The ability of soil to function as a living, dynamic system that sustains plants, animals, and humans over the long term. It emphasizes the biological component of the soil.
Focus	Functional performance of the soil for a specific use (e.g., agriculture, forestry, urban management).	The biological integrity and self-regenerating capacity of the soil ecosystem.
Core Components	- Physical properties (texture, structure, porosity, bulk density) - Chemical properties (pH, nutrient availability, salinity, CEC) - Biological properties (organic matter, microbial activity)	- Organic matter content - Microbial diversity and activity - Ecosystem services (carbon sequestration, biodiversity)
Inherent vs. Dynamic	Involves both inherent (e.g., soil texture, mineral composition) and dynamic properties (e.g., organic matter, structure, nutrient availability).	Focuses primarily on dynamic properties, especially those related to biological processes, such as organic matter, soil microbial activity, and aggregation.
Measurement Approach	- Specific indicators based on soil's current condition and its ability to perform a designated function. - Often measured for short-term productivity.	- Holistic indicators that assess long-term sustainability, focusing on biological processes and regenerative capacities. - Measured for long-term ecosystem health.
Indicators	- Physical: Texture, bulk density, porosity, infiltration rates - Chemical: pH, nutrient levels, salinity, cation exchange capacity (CEC) - Biological: Organic matter content, microbial biomass	- Biological: Soil respiration, microbial activity, earthworm populations, enzyme activity - Organic: Total organic carbon (TOC), humus content - Ecological: Aggregate stability, root development

Temporal Scale	Typically focuses on **short-term** functional outcomes, such as improving crop yields or soil water retention for the current growing season.	Concerned with **long-term** sustainability, resilience to environmental stress, and the capacity for regeneration.
Ecosystem Perspective	Evaluates the soil's capacity to perform specific functions within a given ecosystem but may not address overall soil vitality or resilience.	Evaluates the overall vitality of the soil ecosystem, emphasizing balance and resilience in biological, chemical, and physical interactions.
Management	- Involves specific interventions to improve soil's ability to perform designated functions. - Practices can include fertilization, tillage, irrigation adjustments, and erosion control.	- Emphasizes sustainable, regenerative practices, such as conservation tillage, organic matter addition, cover cropping, and biodiversity enhancement.
Restoration Approach	Restoring soil quality typically involves interventions like adjusting pH, improving drainage, or adding nutrients.	Restoring soil health focuses on fostering biological activity, organic matter buildup, and sustainable nutrient cycling.
Context-Specific	Soil quality is often evaluated in relation to a specific land use or function (e.g., agriculture, urban development).	Soil health is a broader concept and applies across different land uses, emphasizing long-term ecological sustainability.
Short-term vs. Long-term	Primarily concerned with short-term outcomes, such as enhancing soil performance for a single growing season.	Focuses on long-term ecosystem sustainability, resilience, and the soil's ability to recover and adapt over time.
Examples	- High soil quality for agriculture means high crop yields, good nutrient availability, and water retention for the current season.	- High soil health means the soil can sustain plant growth, resist erosion, retain water, and support diverse soil organisms over the long term, with minimal external inputs.

Conclusion

Soil quality and soil health are fundamental to sustainable agriculture, ecosystem stability, and food security. Healthy soils support plant growth, enhance water filtration, sequester carbon, and sustain biodiversity, while degraded soils lead to reduced productivity, erosion, and environmental harm. Maintaining soil health requires sustainable practices such as crop rotation, organic amendments, reduced tillage, and agroforestry. By prioritizing soil conservation and restoration, we can ensure long-term agricultural resilience, mitigate climate change, and protect ecosystems for future generations. Investing in soil health is not just an agricultural imperative but a global necessity for a sustainable future.

References

Amacher, M. C., O'Neil, K. P., & Perry, C. H. (2007). Soil vital signs: a new soil quality index (SQI) for assessing forest soil health. Res. Pap. RMRS-RP-65. Fort Collins, CO: US Department of Agriculture, Forest Service, Rocky Mountain Research Station. 12 p., 65.

Doran, J. W., & Zeiss, M. R. (2000). Soil health and sustainability: managing the biotic component of soil quality. Applied soil ecology, 15(1), 3-11.

Doran, J. W., Jones, A. J., Arshad, M. A., & Gilley, J. E. (2018). Determinants of soil quality and health. In Soil quality and soil erosion (pp. 17-36). CRC Press.

Guo, M. (2021). Soil health assessment and management: Recent development in science and practices. Soil Systems, 5(4), 61.

Harris, R. F., & Bezdicek, D. F. (1994). Descriptive aspects of soil quality/health. Defining soil quality for a sustainable environment, 35, 23-35.

Karlen, D. L., Eash, N. S., & Unger, P. W. (1992). Soil and crop management effects on soil quality indicators. American Journal of Alternative Agriculture, 7(1-2), 48-55.

Karlen, D. L., Mausbach, M. J., Doran, J. W., Cline, R. G., Harris, R. F., & Schuman, G. E. (1997). Soil quality: a concept, definition, and framework for evaluation (a guest editorial). Soil Science Society of America Journal, 61(1), 4-10.

Lehmann, J., Bossio, D. A., Kögel-Knabner, I., & Rillig, M. C. (2020). The concept and future prospects of soil health. Nature Reviews Earth & Environment, 1(10), 544-553.

M. Tahat, M., M. Alananbeh, K., A. Othman, Y., & I. Leskovar, D. (2020). Soil health and sustainable agriculture. Sustainability, 12(12), 4859.

M. Tahat, M., M. Alananbeh, K., A. Othman, Y., & I. Leskovar, D. (2020). Soil health and sustainable agriculture. Sustainability, 12(12), 4859.

Maharjan, B., Das, S., & Acharya, B. S. (2020). Soil Health Gap: A concept to establish a benchmark for soil health management. Global Ecology and Conservation, 23, e01116.

Maikhuri, R. K., & Rao, K. S. (2012). Soil quality and soil health: A review. Int. J. Ecol. Environ. Sci, 38(1), 19-37.

Purakayastha, T. J., Pathak, H., Kumari, S., Biswas, S., Chakrabarty, B., Padaria, R. N., ... & Singh, A. (2019). Soil health card development for efficient soil management in Haryana, India. Soil and Tillage Research, 191, 294-305.

Romig, D. E., Garlynd, M. J., Harris, R. F., & McSweeney, K. (1995). How farmers assess soil health and quality. Journal of soil and water conservation, 50(3), 229-236.

Schloter, M., Dilly, O., & Munch, J. C. (2003). Indicators for evaluating soil quality. Agriculture, Ecosystems & Environment, 98(1-3), 255-262.

Turmel, M. S., Speratti, A., Baudron, F., Verhulst, N., & Govaerts, B. (2015). Crop residue management and soil health: A systems analysis. Agricultural Systems, 134, 6-16.

3

Salt Affected Soil and Reclamation

Introduction

Salt-affected soils (SAS) can be found across different latitudes, exhibiting variations in morphology, biology, and chemical, physical, and physicochemical properties (Gupta and Gupta, 2017). Despite their diversity, SAS share a common characteristic: the dominant role of soil solution electrolytes in the soil formation process, leading to properties that create unfavorable conditions for the normal growth of most plants (Szabolcs, 1989). In cases where SAS develop due to natural geomorphological processes, certain plant species, known as halophytes, have evolved specialized physiological adaptations to survive in these environments. Halophytes are capable of thriving in highly saline habitats by effectively managing the stress caused by excessive salt. Plants growing in such conditions employ various adaptive mechanisms to tolerate and flourish despite the salinity challenge.

Salt-affected soils (SAS) require specialized management strategies to mitigate the challenges posed by high salinity levels. Plants and organisms in these ecosystems have evolved various adaptations to manage salt levels, such as developing specialized root structures, eliminating or containing excess salt, and efficiently absorbing water and nutrients (Flowers and Colmer, 2015). Despite their typically low alpha-diversity, SAS ecosystems contribute significantly to beta-diversity due to the specialization of organisms that thrive in such challenging environments.

However, many of these high-value ecological systems have been severely degraded worldwide as a result of various political and economic decisions. For example, mangroves in countries like Thailand, Vietnam, Mexico, and Ecuador have been destroyed to make way for shrimp farming (FAO, 2007). Sabkhas around the Mediterranean coast have suffered from overexploitation of the aquifers that sustain them, and the reduction of water flow to closed drainage basins, such as in the case of the Aral Sea, has occurred due to water being diverted for irrigation, especially for cotton farming.

Further degradation is seen in the transformation of SAS wetlands into agricultural fields through drainage, a practice observed in many regions (Mitsch and Gosselink, 2007). These ecosystems, once considered wastelands due to their lack of direct economic or agricultural productivity, have been undervalued and transformed, resulting in significant loss of biodiversity and ecological function.

Distribution

In India estimated that about 7 million hectares of land are salt affected i.e. saline and alkaline soils. State wise distribution of salt affected soils in India is given in table 1 (CSSRI, Karnal).

Sr. No.	State	Saline soils (ha)	Alkali soils (ha)	Coastal saline soil (ha)	Total (ha)
1	Andhra Pradesh	0	196609	77598	274207
2	A & N islands	0	0	77000	77000
3	Bihar	47301	105852	0	153153
4	Gujarat	1218255	541430	462315	2222000
5	Haryana	49157	183399	0	232556
6	J & K*	0	17500	0	17500
7	Karnataka	1307	148136	586	150029
8	Kerala	0	0	20000	20000
9	Maharashtra	177093	422670	6996	606759
10	Madhya Pradesh	0	139720	0	139720
11	Orissa	0	0	147138	147138
12	Punjab	0	151717	0	151717
13	Rajasthan	195571	179371	0	374942
14	Tamil Nadu	0	354784	13231	368015
15	Uttar Pradesh	21989	1346971	0	1368960
16	West Bengal	0	0	441272	441272
	Total	**1710673**	**3788159**	**1246136**	**6744968**

Sources of Salt-Affected Soils

Salt-affected soils develop as a result of various natural and human-induced processes that lead to the accumulation of soluble salts in the soil profile. These salts include chlorides, sulfates, carbonates, and bicarbonates of sodium, calcium, and magnesium. The sources of salt-affected soils can be broadly categorized into **natural processes** and **anthropogenic activities**.

1. Natural Sources

Natural processes contribute significantly to the formation of salt-affected soils, particularly in arid and semi-arid regions where evaporation exceeds precipitation. The primary natural sources include:

a) **Weathering of Parent Material:** Many soils inherit their saline or sodic properties from the parent rocks or sediments from which they are formed. The weathering of minerals such as halite (NaCl), gypsum ($CaSO_4$), and other salt-bearing rocks releases soluble salts into the soil. In coastal areas, this is especially prominent due to the proximity to marine deposits that naturally contain high levels of salts.

b) **Capillary Rise of Groundwater:** In regions with high water tables, especially those with saline groundwater, salts from deeper soil layers can move upwards through capillary action. As water evaporates from the soil surface, it leaves behind the dissolved salts, resulting in a gradual increase in salinity near the surface. This is particularly common in areas where groundwater is shallow and highly saline.

c) **Sea Water Intrusion:** In coastal regions, sea water intrusion into the soil can occur due to natural processes or human activities, such as over-extraction of groundwater. When freshwater is depleted, saline water from the sea moves into the aquifers, leading to the accumulation of salts in the soil. This process is exacerbated by rising sea levels due to climate change.

d) **Atmospheric Deposition:** Salts can also be deposited on the soil surface through precipitation or wind. In coastal areas, salt spray from the ocean can be transported inland by wind and rain, contributing to soil salinity. Similarly, in regions close to salt flats or salt lakes, wind can carry fine particles of salt and deposit them onto agricultural lands.

e) **Evaporation in Arid Regions:** In arid and semi-arid regions, the high rate of evaporation compared to precipitation leads to the accumulation of salts in the upper soil layers. As water from the soil surface evaporates, the salts dissolved in the soil solution are left behind, leading to increased soil salinity over time.

2. Human-Induced Sources

Anthropogenic activities are a major cause of soil salinization, often resulting from improper land and water management practices. Key contributors include:

a) **Irrigation Practice:** Irrigation is the most significant human-induced factor leading to the formation of salt-affected soils. In areas with poor drainage, the excessive application of irrigation water without adequate leaching causes salts to accumulate in the root zone. Irrigation water often contains small amounts of dissolved salts, and when large volumes of water are applied over time, these salts concentrate in the soil. Moreover, in areas with high evaporation rates, water used for irrigation evaporates quickly, leaving behind salts in the soil.

b) **Use of Saline Water for Irrigation:** In water-scarce regions, farmers may resort to using saline or marginal quality water for irrigation. Over time, the continuous use of such water adds substantial amounts of salts to the soil, increasing its salinity. This is particularly problematic when freshwater sources are limited, and groundwater is heavily extracted, leading to a higher concentration of salts in the remaining water.

c) **Poor Drainage Systems:** Inadequate drainage is a major factor that contributes to salt accumulation in irrigated lands. Without proper drainage, water containing dissolved salts cannot be flushed out of the soil profile, leading to waterlogging and the subsequent rise of saline groundwater. This issue is prevalent in many agricultural areas where drainage systems are either lacking or insufficient to handle the volume of irrigation water applied.

d) **Overuse of Chemical Fertilizers:** The excessive use of chemical fertilizers, especially those containing sodium, can lead to the accumulation of salts in the soil. This is particularly true in intensively cultivated areas where high amounts of synthetic fertilizers are applied repeatedly without proper management. Over time, the salts from these fertilizers build up in the soil, reducing its fertility and structure.

e) **Industrial and Urban Activities:** Urbanization and industrial activities can also contribute to soil salinization. Industrial effluents, wastewater, and runoff from urban areas may contain high levels of salts that are deposited in nearby agricultural fields. Additionally, road salt used for de-icing can enter the soil through runoff, further contributing to salinity issues in areas close to urban centers.

f) **Land Clearing and Deforestation:** The clearing of natural vegetation for agriculture or other development projects can alter the hydrological balance of the area, leading to changes in groundwater levels and increased salinity. Without deep-rooted vegetation to absorb water, the water table rises, bringing salts closer to the surface. This problem is particularly acute in regions where forests or natural wetlands are cleared for agricultural expansion.

3. Global Examples of Salt-Affected Soils

Many regions around the world are facing the consequences of salinity and sodicity due to the factors mentioned above:

- **The Aral Sea Basin** in Central Asia is a classic example where large-scale irrigation projects for cotton farming led to a dramatic rise in soil salinity. The diversion of water from the rivers that fed the Aral Sea caused the lake to shrink, and the surrounding lands became highly saline.

- **India's Indo-Gangetic Plains** suffer from salinity and sodicity due to improper irrigation practices and poor drainage systems. The overuse of groundwater and inadequate leaching have led to widespread soil degradation in this fertile agricultural region.
- **Australia's Murray-Darling Basin** has experienced severe salinization due to irrigation practices and the use of saline groundwater. Poor drainage and deforestation have also contributed to the rise in soil salinity in this area.

Classification of Salt-Affected Soils

Salt-affected soils are broadly classified into three main types: **saline soils**, **sodic soils**, and **saline-sodic soils**. This classification is based on their chemical properties, such as electrical conductivity (EC), exchangeable sodium percentage (ESP), and pH. Each type of salt-affected soil has distinct characteristics and requires specific management practices.

1. Saline Soils (Solonchak)

- **Electrical Conductivity (EC):** Greater than 4 dS/m (4 mmhos/cm).
- **Exchangeable Sodium Percentage (ESP):** Less than 15.
- **pH:** Usually less than 8.5.
- **Characteristics:** Saline soils contain high concentrations of soluble salts (e.g., chlorides and sulfates of sodium, calcium, and magnesium), which affect plant growth by reducing the availability of water to plant roots. They typically have a white salt crust on the surface, hence the former name "white alkali soils."
- **Effect on Plants:** Plants grown in saline soils experience osmotic stress, which limits their ability to absorb water, even when the soil appears moist. This leads to stunted growth and reduced crop yields.

2. Sodic Soils (Solonetz)

- **Electrical Conductivity (EC):** Less than 4 dS/m (4 mmhos/cm).
- **Exchangeable Sodium Percentage (ESP):** Greater than 15.
- **pH:** Usually between 8.5 and 10.0.
- **Characteristics:** Sodic soils contain excessive sodium ions (Na^+), which displace calcium and magnesium in the soil, causing soil structure deterioration. These soils are poorly drained and prone to waterlogging, with a compact, hard crust that restricts root penetration and water

infiltration. Historically, they were referred to as "black alkali soils" due to the presence of dark organic stains on the soil surface.

- **Effect on Plants:** High sodium levels in sodic soils lead to poor soil structure, resulting in restricted root growth and poor plant health. The soil's alkalinity can also limit nutrient availability.

3. **Saline-Sodic Soils**
   - **Electrical Conductivity (EC):** Greater than 4 dS/m (4 mmhos/cm).
   - **Exchangeable Sodium Percentage (ESP):** Greater than 15.
   - **pH:** Usually between 8.5 and 9.0.
   - **Characteristics:** Saline-sodic soils are a combination of saline and sodic properties. They contain both high levels of soluble salts and significant amounts of exchangeable sodium. The salts help mitigate some of the negative effects of sodium by maintaining soil structure, but once salts are leached out, these soils often become sodic, leading to structural degradation.
   - **Effect on Plants:** These soils present a dual challenge of salinity and sodium toxicity, causing both osmotic stress and poor soil structure. As a result, plant growth is often severely restricted, and remediation requires careful management of both salts and sodium levels.

4. Degraded Alkali or Sodic Soils

Degraded alkali or sodic soils form when saline-sodic soils undergo extensive leaching without the presence of calcium or magnesium. During this process, exchangeable sodium is gradually replaced by hydrogen, leading to a slight increase in soil acidity and causing the soil structure to become unstable. These soils are typically referred to as degraded alkali or sodic soils. As a result of leaching, the surface horizon becomes acidic. Sodium hydroxide (NaOH) then reacts with carbon dioxide (CO_2) from the soil, forming sodium carbonate (Na_2CO_3):

In the sub-surface horizon, this reaction results in an alkali soil. Sodium carbonate dissolves organic matter (humus), which is then deposited in the lower soil layers, giving them a dark, black color. However, the H-clay formed through this process is unstable under alkaline conditions. The breakdown of H-clay in such an environment is known as solodization.

Table 1: Important characteristics of salt affected soils

Characteristics	**Saline soil**	**Non-saline alkali soil**	**Saline-alkali soil**	**Degraded alkali soil**
Content of soil	Excess soluble salts of sodium	Absent of soluble salt and presence of excess exchangeable sodium on the soil complex.	Thee soils are both saline and non-saline alkali soil.	Hydrogen (H^+) ions in the upper layer and sodium (Na^+) in the lower layer
Exchangeable calcium/sodium	Exchangeable calcium	Exchangeable sodium	--	--
Colour	White	Black	--	Black in lower layer
Dominant salts	Sulphate ($SO4^{2-}$) chloride (Cl^-) and nitrate ($NO3^-$) of sodium	Sodium carbonate (Na2CO3)	--	Sodium carbonate (Na2CO3) lower layer
Sodium adsorption ratio (SAR)	Less than 13	More than 13	More than 13	< 13 in the surface and > 13 in the lower horizon
Exchangeable sodium percent-tage (ESP)	Less than 15 of total CEC	More than 15 of total CEC	More than 15 of total CEC	Usually more than 15 of total CEC
Soil pH	Less than 8.5	8.5-10.0	More than 8.5	pH about 6.0 in the surface soil and pH 8.5 in the lower layer constituting the main soil body
Physical condition of the soil	Flocculated condition, permeable to water and air. Soil structure optimum	Deflocculated condition, permeability to water and air is poor. Very poor soil structure	Flocculated deflocculated depending upon the presence of sodium salts and Na-clay	Compact low infiltration & permeability. It develops columnar structure
Morphological character	White crust on the soil surface	Due to presence of high amount of exch. Na^+ and high pH clay	--	--

		colloids get dispersed & migrated downward through the soil profile.		
Organic matter	Slightly less than normal soils	Very low due to the presence of sodium carbonate (Na_2CO_3)	Variable	Low
Total soluble salt content	More than 0.1 percent.	Less than 0.1 percent	More than 0.1 percent	Less than 0.1 percent
EC of the saturation extract	>4 dSm^{-1} or >4 mmhos cm^{-1}	>4 dSm^{-1} or >4 mmhos cm^{-1}	>4 dSm^{-1} or >4 mmhos cm^{-1}	>4 dSm^{-1} or >4 mmhos cm^{-1}

Problems of Salt-Affected Soils

Salt-affected soils, including saline and alkali (sodic) soils, pose significant challenges to agriculture due to their adverse effects on plant growth and soil structure. The following outlines the problems associated with both saline and alkali soils.

Saline Soils

Saline soils contain high concentrations of soluble salts, which negatively impact plant growth in several ways:

1. **Barren but Potentially Productive**: While saline soils are often barren, they hold the potential for productivity if properly managed and reclaimed.
2. **High Wilting Coefficient**: The wilting coefficient of saline soils is elevated, meaning plants wilt quickly due to the difficulty in extracting moisture from the soil.
3. **Low Soil Moisture Availability**: Saline soils have reduced available moisture for plants, which limits their ability to grow and thrive.
4. **Impaired Water and Nutrient Absorption**: The presence of excessive salts disrupts the ability of plant roots to absorb water and essential nutrients from the soil.
5. **Increased Osmotic Pressure**: Excess salts in the soil solution raise the osmotic pressure, making it harder for plant roots to extract water. This osmotic imbalance can cause **exoosmosis**, where water is drawn out of the plants, leading to dehydration and even death during dry periods.

6. **Energy Expenditure by Plants**: Due to high salt concentrations, plants must expend more energy to absorb water and exclude salts from metabolically active sites, which diminishes growth. Additionally, important nutrient elements become unavailable to plants, further inhibiting growth.
7. **Salt Toxicity**: High concentrations of soluble salts can lead to direct toxic effects on plants, such as root injury, reduced seed germination, and overall poor plant health.

Alkali or Sodic Soils

Alkali or sodic soils, characterized by excessive exchangeable sodium, cause both physical and chemical degradation of the soil.

1. **Dispersion of Soil Colloids**: Sodium ions (Na^+) cause the dispersion of soil colloids, especially clay and organic matter, which leads to deflocculation. This loss of soil structure results in the development of compact, poorly structured soil, reducing its suitability for plant growth.
2. **Negative Impact on Physical Properties**: The dispersed and compact soil has reduced aeration, hydraulic conductivity, and drainage. Microbial activity is also inhibited, affecting the soil's overall health and fertility.
3. **Caustic Influence**: High levels of sodium carbonate (Na_2CO_3) and sodium bicarbonate ($NaHCO_3$) in sodic soils create a caustic environment, which further degrades soil quality and plant health.
4. **High Hydroxyl (OH^-) Ion Concentration**: Excessive hydroxyl ions in sodic soils raise the soil pH to harmful levels (10.5 or higher), leading to direct damage to plants, including stunted growth and poor development.
5. **Specific Ion Effects**: Excess sodium can induce deficiencies in calcium (Ca^{2+}) and magnesium (Mg^{2+}) due to several factors:
   - Sodium is loosely held in the soil's exchange complex, so it is preferentially released into the soil solution, displacing other essential cations.
   - High pH in sodic soils causes the precipitation of calcium and magnesium as insoluble carbonates (e.g., $CaCO_3$), reducing their availability to plants.
   - Sodium competitively excludes calcium and magnesium from being absorbed by plant roots.
6. **Nutrient Availability**: The high pH of alkali or sodic soils decreases the availability of many essential plant nutrients, including phosphorus (P),

nitrogen (N), calcium (Ca), magnesium (Mg), iron (Fe), copper (Cu), and zinc (Zn), leading to nutrient deficiencies and poor crop growth.

Reclamation of Salt-Affected Soils

Salt-affected soils, which include saline, sodic, and saline-sodic types, are a major challenge for agriculture in arid and semi-arid regions. These soils degrade crop yields, harm soil structure, and reduce land productivity. The reclamation of salt-affected soils requires a combination of physical, chemical, biological, and management interventions aimed at improving soil health and restoring fertility. This chapter discusses the detailed processes and strategies for reclaiming salt-affected soils.

1. Leaching of Soluble Salts

Leaching is the primary method for reclaiming saline soils, as it involves flushing out the excess salts from the root zone by applying large amounts of good-quality water. The success of leaching depends on several factors:

- **Water Quality**: The water used for leaching should have a low concentration of salts to avoid adding more salts to the soil.
- **Water Quantity**: Adequate volumes of water must be applied to dissolve and remove salts from the soil profile.
- **Drainage**: Efficient drainage is crucial to ensure that the salts are carried away from the root zone rather than accumulating at lower soil depths. A well-designed drainage system, either surface or subsurface, ensures the removal of leached salts.

The process of leaching is particularly effective for saline soils where salts are soluble and can be easily removed with irrigation. However, for sodic and saline-sodic soils, leaching alone is insufficient due to the poor soil structure caused by high sodium levels.

2. Application of Amendments

For sodic and saline-sodic soils, the use of chemical amendments is necessary to replace sodium ions with calcium, thereby improving soil structure and permeability. Common amendments include:

Gypsum (Calcium Sulfate - $CaSO_4$):

Gypsum is the most commonly used amendment for sodic soils. It provides calcium ions, which displace sodium from the soil's exchange complex. The reaction improves soil structure and permeability, facilitating better leaching of sodium and other salts. The process can be summarized as:

Equetion

The displaced sodium is then leached out of the soil with water, and the structure of the soil is improved, allowing for better water infiltration and root growth.

Lime (Calcium Carbonate - $CaCO_3$)

In cases where sodic soils are acidic, lime can be used to raise the soil pH and improve calcium availability. Lime reacts slowly in the soil and helps displace sodium over time. Lime is less effective in highly alkaline conditions, where gypsum is preferred.

c. Sulfur and Sulfuric Acid

For soils that are highly alkaline (high pH), sulfur or sulfuric acid can be applied to lower the pH. Sulfur, when oxidized by soil bacteria, forms sulfuric acid, which dissolves calcium carbonate in the soil and increases the availability of calcium. This calcium can then exchange with sodium to improve soil structure.

3. Improvement of Drainage Systems

Efficient drainage is critical to the successful reclamation of salt-affected soils. Without proper drainage, salts that are leached out of the root zone can accumulate in lower soil layers or create waterlogging problems, further exacerbating salinity. The following drainage systems are commonly used:

Subsurface Drainage

Subsurface drains, such as tile drains, are installed below the soil surface to collect and remove excess water and salts. These drains are highly effective in removing deep-rooted salts and preventing waterlogging in irrigated fields. Subsurface drainage systems require careful design and maintenance but offer a long-term solution to salinity problems.

Surface Drainage

Surface drainage involves reshaping the field to allow water to flow away from the soil surface. This prevents water from stagnating and causing salt accumulation at the surface. Proper land leveling and the construction of drainage channels ensure that excess water is quickly removed, especially in areas with high rainfall or irrigation.

4. Biological Reclamation Methods

Biological methods can play a critical role in the reclamation of salt-affected soils, especially in regions with limited resources for chemical or physical reclamation techniques. Key biological methods include:

Salt-Tolerant Crops and Plants

Planting salt-tolerant crops during the early stages of reclamation can help restore soil fertility. Certain crops, such as barley, wheatgrass, and sorghum, can survive in saline conditions and help improve soil structure by increasing organic matter and promoting microbial activity. Over time, these plants can lower the salt content in the soil through natural leaching processes.

Green Manuring

Green manuring involves growing leguminous crops, such as Sesbania, in salt-affected soils and then incorporating the plant biomass into the soil. This increases soil organic matter, improves soil structure, and enhances microbial activity, which can aid in the reclamation process. Organic matter also helps to retain soil moisture, reducing salt accumulation at the surface.

Use of Microbial Inoculants

Certain salt-tolerant bacteria and fungi can be introduced into salt-affected soils to enhance plant growth and nutrient uptake. These microorganisms can improve soil structure, facilitate the leaching of salts, and increase the availability of essential nutrients to plants.

5. Irrigation Management

Efficient irrigation practices are essential for preventing further salt accumulation and promoting the reclamation of salt-affected soils. Key irrigation management practices include:

Controlled Irrigation

Water should be applied in a controlled and measured manner to prevent over-irrigation and waterlogging, which can lead to the rise of saline groundwater. Irrigation scheduling based on crop water requirements helps minimize water wastage and reduces the risk of secondary salinization.

b. Use of Drip and Sprinkler Irrigation

Drip and sprinkler irrigation systems deliver water directly to the root zone, reducing evaporation losses and preventing salt accumulation on the soil surface. These methods are particularly effective in water-scarce regions and help maintain optimal soil moisture levels for plant growth.

c. Alternate Furrow Irrigation

In this technique, only every other furrow is irrigated, allowing salts to be flushed from the non-irrigated furrows. This helps reduce salt buildup while ensuring that plants receive sufficient water.

6. Mulching

Mulching involves covering the soil surface with organic or synthetic materials, such as straw, compost, or plastic. Mulching reduces evaporation, which is a major cause of salt accumulation in arid regions. Organic mulches also improve soil structure and add organic matter, which enhances soil fertility and moisture retention.

7. Agroforestry and Crop Rotation

Agroforestry practices, which integrate trees and shrubs with crops, can improve the reclamation of salt-affected soils. Deep-rooted trees, such as **Eucalyptus** and **Acacia**, help lower the water table and reduce the risk of salinization. Crop rotation with salt-tolerant species also improves soil structure and enhances soil fertility over time.

8. Desalinization of Irrigation Water

In regions with limited freshwater resources, the use of saline water for irrigation can worsen soil salinity problems. Desalinization technologies, such as reverse osmosis and electrodialysis, can be used to purify irrigation water and reduce the salt load applied to the soil. While these technologies are expensive, they are increasingly being adopted in water-scarce regions to address salinity issues.

Conclusion

Salt-affected soils pose a significant challenge to global agriculture, reducing soil fertility, crop productivity, and ecosystem sustainability. These soils, categorized as saline, sodic, and saline-sodic, require appropriate reclamation strategies tailored to their specific conditions. Effective reclamation involves a combination of physical, chemical, and biological methods, including leaching, gypsum application, organic amendments, and the cultivation of salt-tolerant crops. Sustainable water management practices, such as efficient irrigation and drainage systems, are crucial in preventing secondary salinization. The successful reclamation of salt-affected soils not only enhances agricultural output but also contributes to environmental conservation and food security. Continued research, technological advancements, and farmer education are essential to improving reclamation techniques and ensuring long-term soil health. By adopting integrated soil management practices, we can mitigate the adverse effects of soil salinity and promote sustainable land use for future generations.

References

Abrol, I. P., Yadav, J. S. P., & Massoud, F. I. (1988). Salt-affected soils and their management (Vol. 39). Food & Agriculture Org..

Chen, X., Yang, J., She, D., Chen, W., Wu, J., Wang, Y., ... & Souza, E. R. D. (2025). Monitoring, Reclamation and Management of Salt-Affected Lands. Water, 17(6), 813.

Chhabra, R. (2004). Classification of salt-affected soils. Arid Land Research and Management, 19(1), 61-79.

Ezeaku, P. I., Ene, J., & Shehu, J. A. (2015). Application of different reclamation methods on salt affected soils for crop production.

Fagodiya, R. K., Malyan, S. K., Singh, D., Kumar, A., Yadav, R. K., Sharma, P. C., & Pathak, H. (2022). Greenhouse gas emissions from salt-affected soils: Mechanistic understanding of interplay factors and reclamation approaches. Sustainability, 14(19), 11876.

Fullen, M. A., & Catt, J. A. (2014). Soil management: problems and solutions. Routledge.

Gunarathne, V., Senadheera, J. A. I., Gunarathne, U., Almaroai, Y. A., & Vithanage, M. (2020). Reclamation of salt-affected soils. In Soil and Groundwater Remediation Technologies (pp. 183-199). CRC Press.

Gupta, R. K., & Abrol, I. P. (1990). Salt-affected soils: their reclamation and management for crop production. Advances in Soil Science: Soil Degradation Volume 11, 223-288.

Gupta, S. K., & Gupta, I. C. (2014). Salt affected soils: reclamation and management. Scientific Publishers.

Kaledhonkar, M. J., Meena, B. L., & Sharma, P. C. (2019). Reclamation and nutrient management for salt-affected soils. Indian Journal of Fertilisers, 15(5), 566-575.

Keren, R. (2023). Reclamation of sodic-affected soils. In Soil erosion, conservation, and rehabilitation (pp. 353-374). CRC Press.

Kumar, R., Singh, A., Bhardwaj, A. K., Kumar, A., Yadav, R. K., & Sharma, P. C. (2022). Reclamation of salt-affected soils in India: Progress, emerging challenges, and future strategies. Land Degradation & Development, 33(13), 2169-2180.

Osman, K. T. (2018). Management of soil problems (pp. 255-298). Cham, Switzerland: Springer International Publishing.

Oster, J. D., Shainberg, I., & Abrol, I. P. (1999). Reclamation of salt-affected soils. Agricultural drainage, 38, 659-691.

Oster, J. D., Shainberg, I., & Abrol, I. P. (2023). Reclamation of salt-affected soil. In Soil erosion, conservation, and rehabilitation (pp. 315-352). CRC Press.

Thampatti, K. M. (2022). Problem Soils: Constraints and Management. CRC Press.

4

Acid Soils and Their Management

Introduction

In India acid soils occur in the high rainfall areas covering about 25 million hectares of land with a pH below 5.5 and 23 million hectares of land with a pH between 5.6 and 6.5. In India, acid soils occur in Assam, Meghalaya, Arunachal Pradesh, Mizoram, Nagaland, NEFA, Manipur, Tripura, West Bengal, Bihar, Uttar Pradesh, Himachal Pradesh, Jammu and Kashmir, MP, Maharashtra, Kerala, Karnataka, Tamil Nadu and Andhra Pradesh. Punjab, Haryana, Rajasthan and Gujarat are the only states in India where acid soils do not occur. Soil acidity is a limiting factor affecting the growth and yield of many crops all over the world. The basic problems concerning chemical properties of more acid soils are, besides acidity itself, the presence of toxic compounds and elements, such as soluble forms of Al, Fe and Mn, nitrites and various toxic organic acids. Aluminium (Al) toxicity is one of the major constraints on crop productivity on acid soils, which occur on up to 40% of the arable lands of the world. Al is the third most abundant element in the earth's crust and is toxic to plants when solubilised into soil solution at acidic pH values. Very few plants can grow well in strong acid soils. Soil acidity below pH value 5.5 is generally injurious to plants. Plant roots are badly affected if the pH value exceeds limits of tolerance for particular crops. High degree of soil acidity (pH 5.0 to 6.5) decreases the availability of plant nutrients particularly phosphorus, calcium, magnesium, molybdenum, iron, manganese, potassium sulphur nitrogen, boron, copper and zinc. It also affects adversely the important microbiological processes, such as nitrogen fixation by *Azotobacter*, *Clostridium* and nodule inhabiting bacteria (*Rhizobia*) of leguminous plants.

Occurrence

Acid soils occupy approximately 60% of the earth land area and are arise under humid climate conditions from carbonaceous less soil forming rocks in all thermal belts of the earth.

- Worldwide – 800 M ha
- India - 100 M ha
- Tamil Nadu - 2.6 M ha (20% of GA)

95% of soils of Assam and 30% of geographical area of Jammu and Kashmir are acidic. In West Bengal, 2.2 M ha, in Himachal Pradesh, 0.33 M ha, in Bihar, 2 Mha and all hill soils of erstwhile Uttar Pradesh come under acid soils. About 80% of soils in Orissa, 88% in Kerala, 45% in Karnataka and 20% in Maharastra are acidic. The laterite zone in Tamil Nadu is covered with acid soil and about 40,000 ha are acidic in Andhra Pradesh.

a) **Area basis:** Assam & NE states (20 m ha) > Jammu & Kashmir (15.5 m ha) > Orissa (12.5 m ha) > Karnataka (9.6 m ha) > Madhya Pradesh & Chattisgarh (8.9 m ha).

b) **Per cent of geographical area:** HP (90%) > Kerala (90%) > Assam & NE states (80%)> Orissa (80%) > Jammu & Kashmir (70%) > Karnataka (50%).

c) **Soil Types:** Soil acidity is mainly seen in Oxisols (laterites) and Alfisols (red soils), Histosols, Spodosols.

Definition of Soil Acidity

A soil with a preponderance of H ions, and probably of Al and its various hydrated forms in proportion to OH ions, specifically with pH less than 6.5 is known as an acid soil."

Soil acidity involves intensity and quantity aspects. The intensity aspect is universally characterized by the measurements of ion activity, expressed as pH. The quantity aspect is characterized, directly or indirectly, by the quantity of alkali required to titrate soil to some arbitrarily established endpoint. Soil acidity is a major problem in relation to plant growth and therefore, acid soils are called a problem soil.

Classification of Acid Soils Based on Magnitude of Acidity

pH Range	Nature of Acidity
< 4.5	Extremely
4.5-5.0	Very strongly acidic
5.1-5.5	Strong acidic
5.6-6.0	Medium acidic
6.1-6.5	Slightly acidic

According to the amount of organic matter, acid organic soils can be classified in to the following two types:

i. **Peat soils:** Peat soils are characterized by presence of poorly degraded organic matter in India, peaty soils occur in Kashmir, Himachal Pradesh, Assam and other hill states.

ii. **Muck soils:** Such soils contain highly degraded organic matter. They have relatively higher pH values then the peaty soils. Thus they are less acidic. Muck soils are also found in Kashmir, Himachal Pradesh, Assam and other hill states

Characteristics of Acid Soils

1. Acid soils have low pH and high proportion of exchangeable H^+ and Al^{3+}.
2. Kaolinite and illite types of clay minerals are dominant in these soils.
3. These soils have low CEC and low base saturation.
4. These soils have high toxic concentration of Al, Fe and Mn and deficiency of Ca and Mg.
5. These soils have nutrients and microbial imbalances.
6. These soils are generally low in available phosphorus and high P_2O_5 fixation capacity.
7. Soil acidity inhibits biological N-fixation.
8. Sensitivity to erosion
9. Low water-holding capacity
10. Low permeability to air, water, and roots (high soil strength)
11. Slow water infiltration rate
12. Sensitivity to compaction by heavy machinery.

Factors Responsible for of Soil Acidity

The acid soils are caused by a number of things. Acid soils are generally created and developed as a result of a variety of variables, including the climate, hydrologic cycle, vegetation, parent rocks, human intervention, and others. In humid areas with frequent, intense rainfall, acid soils are more common. Acid soils are absent from dry areas. The formation of acid soils is caused by the following elements. Acid soils are base unsaturated soils that have enough adsorbed exchangeable H^+ ions to reduce the pH of the soil below 7.0.

1. Leaching due to heavy rainfall

In dry areas, few bases are often present because little water permeates the soil. When rainfall increases and soluble salt content lowers to a low level, gypsum and $CaCO_3$ are eliminated in the order mentioned. The rate of bases being eliminated eventually surpasses the rate of liberation from non-exchangeable forms as precipitation increases. The pH of the soil is lowered and the concentration of H^+ on exchange complex is increased due to the considerable loss of bases caused by heavy rains and leaching.

2. Acid parent material

Acidic parent material such as acidic igneous rocks like granite, sedimentary rocks like sandstone may weather to give rise acid soils.

3. CO_2

The carbon dioxide evolved during organic matter decomposition and root respiration dissolves in water to form weak carbonic acid.

$\mathbf{CO_2 + H_2O \rightarrow H_2CO_3}$

(Carbonic acid)

H_2CO_3 -------------→ $HCO_3 + H^+$

These acidified waters percolate through the soil and cause soil acidity gradually. The percolating waters continuously move small amounts of H+ ions which replaces solubilized basic cations such as Ca^{2+}, Mg^{2+}, K^+ and Na^+. The replaced cations are then leached from the root zone.

3. **Ionization of water:** The water may ionize and contribute H^+ on exchange complex as follows:

   $\mathbf{H_2O \rightarrow HOH \rightarrow H^+ OH^- \rightarrow H^+[X] + Bases + OH^-}$

3. **Contact exchange:** The contact exchange between exchangeable H on root surface and the bases in exchangeable form on soil particle may take as follows:

4. **Soluble acid production:** The decomposition of organic matter in the soil produces many organic as well as inorganic acids. These acids may contribute H^+ on exchange complex.

5. **Use of nitrogenous fertilizers:** Continuous use of nitrogenous fertilizers containing NH_4-N or giving NH_4-N on hydrolysis (i.e. urea) produce various acids in soils e.g. 1 mole of NH_4 in

   NH_4NO_3 gives 2 moles of HNO_3; 1 mole of $(NH_4)_2SO_4$ gives 2 moles of HNO_3 + 1 mole of H_2SO_4; 1 mole of NH_4OH gives 1 mole of HNO_3. Thus, continuous use of such fertilizers will produce acidity in soil.

6. **Oxidation of FeS:** FeS or iron poly sulphide accumulates under anaerobic conditions as a result of reduction of Fe^{3+} and SO_4. Under aerobic conditions, they will be oxidized and will produce H_2SO_4 as follows:

   $\mathbf{4FeS_2 + 15O_2 + 2H_2O \rightarrow 2Fe_2(SO_4)_3 + 2H_2SO_4}$

   Under such conditions, soil pH values of 2 to 4 are frequently observed.

7. **Hydrolysis of Fe^{3+} and Al^{3+} :** The Fe^{3+} and Al^{3+} ions may combine with water and release H+ as follows:

$$\mathbf{Al + H_2O \rightarrow Al(OH)_2{+} + H_+}$$

$$\mathbf{Al(OH)^{2+} + H_2O \rightarrow Al(OH)_2^{+} + H^+}$$

$$\mathbf{Al(OH)_2^{+} + H_2O \rightarrow Al(OH)_3 + H^+}$$

The hydrogen produced may enter on exchange complex.

8. **Acidification from the air:** Industrial exhausts, if contain appreciable amount of SO2 may cause acidity in soil in course of time due to dissolution of SO2 in water (rain) as follows :

   $$\mathbf{SO_2 + H_2O \rightarrow H_2SO_3}$$

   (Rain water) (Sulphuric acid)

   The phenomenon gives acid rain. In soil, the reaction may be as follows:

   $$\mathbf{2H_2SO_3 + O_2 \rightarrow 2H_2SO_4}$$

9. **Crop removal of bases:** Crop removal of Ca^{2+}, Mg^{2+} and K^+ makes the soil more acidic.

Kinds of Soil Acidity

Broadly soil acidity may be of two kind's *viz.,* (i) Active acidity (ii) Potential reserve/exchange acidity. The nature of soil can be illustrated as follows:

Adsorbed H (and Al) ions on soil colloids ↔ Soil solution H (and Al) ions

(Potential/exchange/reserve) (Active acidity)

i) **Active acidity:** Active acidity may be defined as the acidity develops due to hydrogen (H^+) and aluminium ($Al^{3+)}$ ions concentrations of the soil solution. The magnitude of this acidity is limited. The acidity in soil solution is known as active acidity and is measured by pH meter.

ii) **Exchange acidity:** Exchange acidity may be defined as the acidity develops due to adsorbed hydrogen (H^+) and aluminium ($Al^{3+)}$ ions on the soil colloids. The magnitude of this acidity is very high.

However, residual acidity could be added to the total acidity. The acidity that is still present after the active and exchange acidities have been neutralised is known as residual acidity. It is linked to aluminium hydroxyl ions as well as hydrogen and aluminium atoms that are attached in non-exchangeable ways by organic materials and silicate clay. By using the $BaCl_2$ + Triethanol amine reagents to calculate the exchangeable H, passive acidity, also known as reserve acidity, is evaluated. Total acidity, also known as titratable acidity, is the sum of the H ion concentrations in the exchange complex and in solution, both of which can be determined using titration. Acidity in all of its forms is in balanced.

Total acidity: The total acidity is summation of active, exchange and residual acidity. It can be written as,

Total acidity = Active acidity + Exchange acidity + Residual acidity

Therefore, total soil acidity depends on the active, exchange and residual acidity of the soil.

Active acidity – soluble acidity, in the solution

Reserve acidity – adsorbed acidity, on the surface of particles

Problems of Soil Acidity

Problems of soil acidity may be divided into three groups which are as follows:

1. Toxic effects

a) Acid toxicity

Under extremely acidic soil conditions, plants become poisonous due to the higher hydrogen ion concentration. Acid anions and H^+ ions may both be hazardous as part of the acid toxicity.

b) Toxicity of different nutrient elements

Iron and Manganese toxicity

The concentration of iron and manganese ions (Fe^{2+} and Mn^{2+}) in a soil solution is influenced by the soil reaction, or pH, the amount of organic matter, and the degree of soil reduction. The population of soil microorganisms grows as the amount of organic matter in the soil rises, and this leads to a rapid depletion of soil oxygen and soil degradation. The nutritional elements Mn^{4+} and Fe^{3+} convert to Mn^{2+} (manganous manganese) and Fe^{2+} (ferrous iron), respectively, as a result of soil reduction. This raises their concentration to an extremely high level, where toxicity of those elements arises. A physiological disease of rice called as browning disease, which is caused by similar toxic effects, is observed in submerged soils.

Aluminium toxicity

The cations that are present along with Al may have a significant impact on its toxicity. Al's toxicity tends to decline as the concentration of other cations, like calcium, rises. Both upland and lowland soils are susceptible to al toxicity. The toxicity of aluminium in soils has a variety of effects on plant growth.

1. It limits the expansion of the roots.
2. It has an impact on a variety of physiological functions in plants, including cell division, DNA synthesis, and respiration.

3. It limits the plant's ability to absorb and move several key nutrients from the soil, such as phosphorus, calcium, iron, and manganese.
4. Plants begin to wilt as a result.
5. It also suppresses the soil's microbiological activity.

2. Nutrient availability

a) Non-specific effects

It is associated with the inhibition effect of root growth and thereby affects the nutrient availability.

b) Specific effects

i. **Exchangeable bases:** The availability of exchangeable bases has two components: the ion uptake process and the release of bases from the exchangeable form, both of which may be negatively impacted by soil acidity. Exchangeable bases are released preferentially in a fractional exchange as a result of complimentary ion effects. Acidic soils have deficiencies in bases like Ca^{2+} and Mg^{2+}.

ii. **Nutrient imbalances:** It is clear that moderate to strongly acidic soils typically contain larger quantities of soluble iron, aluminium, and manganese. Plants cannot use phosphorus because of the reaction between phosphorus and these ions, which results in insoluble phosphatic compounds. In addition to these, the availability of phosphorus is reduced by the fixing of phosphorus by hydrous oxides of iron and aluminium or by adsorption. Iron, manganese, copper, and zinc are all plentiful in acidic soils, but molybdenum is scarce and inaccessible to plants. The availability of boron may also be reduced in acidic soils with very low pH levels due to adsorption on sesquioxides, iron, and aluminium hydroxy compounds. In an acidic soil with a pH below 5.5, nitrogen, potassium, and sulphur become less readily available.

3. Microbial activity

It is widely known that changes in the soil reaction have an impact on soil organisms. Actinomycetes and bacteria perform better in soils with moderate to high pH levels. When the pH of the soil falls below 5.5, they are unable to exhibit activity. Reduced Azotobacter sp. activity significantly affects nitrogen fixation in acidic soils. In addition to this, soil acidity also prevents symbiotic nitrogen fixation by influencing *Rhizobium sp.* activity. In extremely acidic soils, fungi can thrive and produce a variety of illnesses, including potato blights and tobacco root rot.

Management of acid soils

The management of acid soil should aim at improving the production potential of the soil by either amendments or manipulation of agricultural practices to obtain optimum yield under acidic conditions. However, even within the crop, the performance of varieties also varies under acidic conditions. Acid soils can be managed either by growing crops suitable for a particular soil pH or by ameliorating the soils through application of amendments, either inorganic or organic, which can counteract soil acidity. Besides these, good agronomic management practices are also important to manage soil acidity to enhance the productivity of the crop in acidic environment. Some practices are given here.

A. Liming

Amelioration of soil and sub-soil acidity constitutes an important aspect of acidic soil management. Application of lime and/or alternative liming material along with other management practices are needed to address soil acidity. Apart from increasing yield, application of lime enhances the efficiency of applied fertilizers, improves the effectiveness of some herbicides, protects the environment and increases the net profit of farmers.

$$\mathbf{2H - clay + CaCO_3 \rightarrow Ca - clay + H_2O + CO_2 \uparrow}$$

Liming material: The common liming material used to reclaim acid soils are:

1. **Quick lime (CaO)** which is also known as burnet lime, oxide of lime or simply lime (or calcium oxide). Oxide of lime is more caustic than limestone. Burned lime is produced by heating limestone and dolomite as follow:

   CaCO3 + Heat → CaO + CO2

   CaMg (CO3)2 + Heat → CaO + MgO + 2CO2

2. **Slaked lime [(Ca(OH)$_2$]** or hydrated lime or also known as hydroxide of lime (or calcium hydroxide). It can be produced by adding water to burned lime and is called slaked lime.

   $CaO + H_2O \rightarrow Ca(OH)_2$

   (Burned lime) (Slaked lime)

   It is more caustic than burned lime (CaO). If it is kept open in the moist air, then combination of calcium hydroxide occurs as follows:

   $Ca(OH)_2 + CO_2 \rightarrow CaCO_3 + H_2O$

3. **Calcic lime stone ($CaCO_3$)** also known popularly as agriculture lime or carbonate of lime or ground lime stone.

4. **Dolomitic lime stone ($CaMg(CO_3)_2$)** dolomite form of ground lime stone containing both Calcium and Mg.

5. **Slags:** These are generally three types of slags that are found important:
   a. **Blast furnace slag.** It is a by-product of the manufacture of pig iron. As a liming material, this slag behaves essentially as calcium silicate. The neutralizing value of blast furnace slags ranges from shout 75-90%.
   b. **Basic slag.** It is a by-product of the basic open-hearth method of making steel from pig iron, which in turn is produced from high phosphorus iron ores. The impurities in the iron, including silica and phosphorus are fluxed with lime and the basic slags are produced. Its neutralizing value ranges from 60 to70%.
   c. **Electric furnace slag.** This is produced from the electric furnace reduction of phosphate rock during preparation of elemental phosphorus. This product is largely calcium silicate and is used as a liming material.
   d. **Other liming materials:** Coral shell, chalk, wood ash, press mud, by product material of paper mills, sugar factories, fly ash sludge etc. are considered as liming materials and also used for the amelioration of soil acidity.

Factors affecting liming

1. **Soil texture** – sandy soils require less lime than clays to change soil pH. Because of their higher buffering capacity clays take longer to acidify but need more lime to change the pH
2. **Rate of lime applied** – in the region rates vary from 1.0 to 3.0 tonnes per hectare depending on the soil pH and soil texture. Higher rates can reduce frequency of applications. Even spreading is essential. Caution: High lime application rates can induce manganese deficiency especially on sandy soils
3. **Desired pH range** – the lower the soil pH, the more lime is needed to recover ideal levels. In soils where the sub-surface layer (10-20cm) pH is low the surface pH may need to be maintained at a higher level through liming to enable movement of lime into the sub-surface layer.
4. **Moisture-** Greater the moisture content, more rapid the rate of reaction.
5. **Temperature-** Lime and liming material reacts more rapidly at high temperature than at low temperature.

Benefits of liming

- Reduces soil acidity and improves soil pH, base saturation and CEC
- Increases nutrient availability

- Changes insoluble soil complexes of P and S to more plant available forms
- Encourages the growth of microorganisms such as nitrogen fixing bacteria, nitrifying bacteria and organic matter decomposing bacteria etc.,
- Improves nitrogen fixing by legumes
- Stabilizes the soil structure by forming C-humate complex.
- Reduces Fe, Al and Mn toxicities
- Improves the effectiveness of certain herbicides
- Reduces fungal diseases
- Increases crop yields.

A. Harmful effects of over-liming

One of the most detrimental effects of over-liming is the alteration of the physical properties, rather than the chemical properties in tropical soils. Soil permeability is also known to be affected by over-liming. High infiltration rates and consequent rapid leaching of bases from tropical soils are attributed to highly unstable soil structure and increased binding tendency of ferrous and aluminum oxides in soil particles. Over-liming destabilizes the soil structure, which in turn causes soil aggregates to break apart resulting in reduced permeability and inadequate drainage. The addition of lime of either calcium or magnesium to soil increases the number of small aggregates at the expense of larger ones.

B. Pulp and paper mill effluents

Huge quantities of effluents are generated from pulp and paper, tannery and textile industries that could be used in managing acid soils. Paper Mill Sludge (PMS) has been tested for its suitability and found to be a good and cheap source compared to calcite and dolomite.

Lime sludge is a solid waste produced while converting wood/bamboo chips into pulp in the paper industry. Its major component is $CaCO_3$; it contains low levels of potentially toxic heavy metals and can be a cheap source of amelioration of soil acidity.

C. Phospho-gypsum (PG) or gypsum

Chemically, gypsum ($CaSO_4.2H_2O$) is a neutral salt with no direct effect on soil pH. However, many researchers have shown that phospho-gypsum, a by-product in the production of phosphoric acid from phosphate ore and sulfuric acid, can ameliorate sub-soil acidity and hasten root development.

Phosphogypsum contains 16% sulphur (S), 20-21% calcium (Ca) and 0.2-1.2% phosphorus (P). This is very relevant in rainfed ecosystems where the absorption of water and nutrients is limited due to poor development of the root system. Since Ca in Phosphogypsum can leach down faster compared to lime in light textured soils, sub-soil acidity could be ameliorated with resultant reduction of aluminium toxicity and Ca deficiency.

D. Growing acid tolerant crops

Aluminium toxicity limits crop production in acidic soils, to which soil liming is the answer. However, considering the huge quantities of lime and associated costs involved in amelioration of these soils, growing acid-tolerant crops and cultivars might be a viable alternative. Blueberries, potatoes and watermelons tend to be more acid tolerant than crops like corn, soybean, wheat, alfalfa and clover. There is considerable variability in Al tolerance among plant species, which has enabled breeders to develop Al-tolerant cultivars and study the physiology and biochemistry of Al tolerance in germplasm. Wheat has proven to be a useful candidate in this respect, with up to 10-fold difference in Al tolerance among its genotypes compared to other cereals. Paddy is a good choice because flooding neutralizes the acidity and associated negative effects where water is abundantly available.

Relative tolerance of crops to soil acidity

Crops	Optimum pH range
Cereals	
Maize, sorghum, wheat, barley	6.0-7.5
Millets	5.0-6.5
Rice	4.0-6.0
Oats	5.0-7.7
Legumes	
Field beans, soybean, pea, lentil etc.	5.5-7.0
Groundnut	5.3-6.6
Others	
Sugarcane	6.0-7.5
Cotton	5.0-6.5
Potato	5.0-5.5
Tea	4.0-6.0

Relative tolerance of fruit crops to soil acidity

Slightly tolerant (pH 6.8-6.0)	**Moderately tolerant (pH 6.8-5.5)**	**Highly tolerant (pH 6.8-5.0)**
Mango	Pine apple	Strawberry
Citrus	Orange	Gooseberry
Banana	Litchi	Bael
Guava	Passionfruit	Elephant apple
Papaya	Jackfruit	Plum
Apple		
Peach		
Kiwi		

Relative tolerance of vegetables to soil acidity

Slightly tolerant (pH 6.8-6.0)	**Moderately tolerant (pH 6.8-5.5)**	**Highly tolerant (pH 6.8-5.0)**
Beet	Frenchbean	Potato
Cabbage	Carrot	Sweet potato
Cauliflower	Cucumber	Chow-chow
Broccoli	Brinjal	Watermelon
Celery	Pea	Chicory
Spinach	Pumpkin	Dandelion
Chinese Cabbage	Radish	Yam
Leak	Tomato	
	Garlic	

E. Use of basic fertilizers

$NaNO_3$ and basic slag, etc.

F. Soil and water management

Proper soil and water management checks leaching of bases and enhances decomposition of organic matter.

G. Application of organic amendments

The conventional liming materials such as calcium carbonate and calcium oxide are not only scarce and expensive; their application rate is also very high which ultimately leads to higher production cost per unit area. Consequently, need of the hour is to focus research on development of locally adapted indigenous and sustainable way of managing soil acidity. Organic waste amendments, such as farm yard manure (FYM), green manures, Poultry manure, pig manure, biochar etc., are usually used as an alternative to lime application for the reclamation of acid soil infertility and build-up of depleted soil organic matter. External application of animal manure has been reported to increase soil pH and decline Al toxicity in acid soil, At the same time, animal manure supplies the crops with available N R and K and improves physico-chemical and biological properties of soil.

i. **Poultry manure:** Indiscriminate dumping of poultry manure on the environment have been a source of concern to poultry farmers, environmentalists and the general public. However, the same poultry manure can be utilised as an amendment for the management of acidic soil in tropical and sub-tropical countries. On an average, poultry manure contains 3-5% nitrogen, 1.5-3.5% phosphorus, 1.5-3.0 % potassium and micronutrients at considerable amount. Poultry manure contains high ammonical-N (NH_4-N) and uric acid which is capable of converting a large percentage of N to nitrate-N (NO_3-N) within a week as uric acid metabolizes rapidly to NH_4-N in most soils. Moreover, the use of poultry manure has overtaken the use of other organic manures/ such as pig manure, due to its high content of both macronutrients (N P, and K) and micronutrients (Fe, Mn, Cu and B) besides improving the soil structure, nutrient retentiory aeratiory filtration and water holding capacity of soil. Since poultry manure is cheap source of both macronutrients (N, P, K, Ca, Mg, S) and micronutrients (Fe, Mn, Cu and B) and has the potential to increase the porosity and microbial activity in soil, it can be used as an alternative to compensate for shortage of FYM in NEH region for management of acid soil. Moreover, the application of poultry manure elevates the soil pH mainly due to the presence of substantial amount of $CaCO_3$ in it.

ii. **Pig manure:** Pig farming is one of the major sources of income for small and marginal farmers of NEH region. Pig manure can be considered as an important source of nutrients for crop production due to the presence of N, P, K, Ca, Mg, S, Mn, Cu, Zn, Cl, B, Fe, and Mo. Although pig manure contains relatively lower amounts of N and K bub, it has higher levels of Cu and Zn because these nutrients are added to feed. Although application of pig manure sometimes helps to improve the soil acidity, however, further research is needed to prove its authenticity.

iii. **Green manures:** Green manuring can be defined as the growth of a crop for the specific purpose of incorporating it into soil while green, or soon after maturity with a view to improve the soil and benefit subsequent crops. Green under decomposed material used as manure is called green manure. The most important green manure crops are Sunnhemp, Dhaincha, Clusterbean, Sesbania rostrata etc. Application of green manures in acid soils can also reduce aluminium toxicity and subsequently increase crop yield. Addition of green manure also improves soil structure and increases water holding capacity of soil. Studies indicated that addition of green manure in acid sulphate soil resulted in significant reduction in aluminium concentration. Moreover, peat mixed with green manure and rice straw synergistically promoted complexation and chelation of monomeric aluminium, presumably forming aluminium-organic acid complex in the soil. However, higher application rate of green manures and/or organic manures is practically not feasible thus for increasing soil pH, decreasing concentration of phytotoxic aluminium and reducing lime requirements, combined application of lime and organic manure is advisable for management of acid soil.

iv. **Biochar:** Biochar generally refers to the black carbon or carbon-rich solid product which is produced by pyrolysis of biomass under oxygen free environment, so that it is not subjected to complete combustion. Use of biochar as soil amendment has been receiving significant attention for various reasons such as neutralizing acidity in soil, creating a carbon sink to mitigate global warming, increasing soil water holding capacity, reducing greenhouse gas emissions and stabilizing mobile heavy metals, pesticides and other organic pollutants in soil. Free bases present in biochar such as Ca, Mg and K can be readily released into the soil solution which leads to net increase of soil pH and exchangeable bases and ultimately helps to mitigate the soil acidity. Thus, biochar could be used as an alternative of lime for management of soil acidity in NEH region. However, its production cost and availability restrict its use in agriculture as an amendment.

H. Good agronomic practices

i. **Crop Scheduling:** In order to get good crop growth and productivity, under rainfed conditions, highly responsive crops like cotton, soybean, pigeon pea, groundnut French bean etc. should be grown in the first year of liming. This should be followed by medium responsive crops like maize and wheat in the subsequent seasons. Cultivation of lowest responsive crops like millets, rice, barley, linseed etc. should be done in the last season and/or year prior to commencement of the next liming schedule.

ii. **Crop diversification:** Location specific crop diversification with acid tolerant crops like rice, tea, groundnut, maize etc. should be followed. Crop diversification especially in rainfed uplands through intercropping and in rainfed medium and lowlands through sequence cropping is essential for increased agricultural productivity of acid soil regions.

iii. **Lime application techniques:** Lime placement in soil is highly important in determining reactivity and movement into the sub soil. Liming materials have very limited movement into the soil without incorporation. Hence, lime is usually broadcasted on the soil surface and wherever possible, the added lime should be incorporated in the soil as deeply as possible by a plough or disc harrow. It has been usually suggested that the recommendation of lime should be made in small doses to make it more cost-effective and for better acceptability of liming practice by resource poor farmers. It was also suggested to apply the lime with basal dose of fertilizers (part of N + full dose of P and K) manually or through seed-cum-fertilizer drill at the time of sowing.

iv. **Seed priming:** Seed priming is a pre-sowing treatment which leads to a physiological state that enables seed to germinate more efficiently and ultimately helps in strengthening crop establishment and increase the crop yields under adverse situation.

v. **Integrated nutrient management practice:** Integrated nutrient management (inclusive of lime, organic manure and inorganic fertilizers) is often advocated to boost the crop productivity in acid soils.

vi. **Growing of multi-purpose tree species:** The multi-purpose tree species (MPTs), besides furnishing the multiple outputs like fuel, fodder, timber and other miscellaneous products, help in improvement of soil health and other ecological conditions. Studies revealed that MPTs with greater surface cover, constant leaf litter fall, and extensive root system increased soil organic carbon content porosity, aggregate stability and available soil moisture by 33.2%, 10.9%, 24.07%, and 33.2%, respectively, and simultaneously reduced bulk density and erosion ratio by 15.9% and 39.5%, respectively in acid soils North-East India.

vii. **Adoption of agro-forestry system:** Agro-forestry system (AFS) has become an alternative approach of integrated land management system not only for renewable resource production but also for improving soil health and ecological conditions. A study indicated that in all the AF$ significant (1.17-1.65- fold) increase in organic carbon was found as compared to initial status, the maximum contribution being by silvi-horti-pastoral AFS. The same system also registered 43.2% higher exchangeable Al compared to natural forest and consequently a maximum decrease of 0.50 units in pH. Moreover, the exchangeable Ca, Mg, Na, K and Al and available N and P content were higher in all the systems compared to natural forest. It is suggested that AFS could be used as a potential alternative approach for management of acid soil in NEH region of India.

viii. Crop residues: Crop residues can improve soil structure, increase organic matter content in the soil. Organic matter addition as crop residues in rotation or as intercrops serves as components of acid soil management systems. Moreover, the organic 'matter to the soil reduces the concentration of Al and Fe and increases the availability of P in acid soils. Studies indicated that addition of grass residues substantially increased the maize yield and the concentrations of exchangeable (Ca and Mg) and extractable P in acid soils.

Conclusion

Acid soils pose significant challenges to agricultural productivity due to their low pH, aluminum toxicity, and nutrient deficiencies. However, through effective management practices such as liming, organic matter incorporation, balanced fertilization, and the cultivation of acid-tolerant crops, their fertility can be improved. Sustainable soil management not only enhances crop yields but also ensures long-term soil health, making it essential for food security and environmental conservation in affected regions. Adopting integrated approaches tailored to local conditions is key to maximizing agricultural potential in acid soil areas.

References

Abebe, M. (2007). Nature and management of acid soils in Ethiopia.

Chintala, R., Mollinedo, J., Schumacher, T. E., Malo, D. D., & Julson, J. L. (2014). Effect of biochar on chemical properties of acidic soil. Archives of Agronomy and Soil Science, 60(3), 393-404.

Craswell, E. T., & Pushparajah, E. (1989). Management of acid soils in the humid tropics of Asia. Australian Centre for International Agricultural Research/International Board for Soil Research and Management.

Dash, A. K., Singh, C., & Varma, S. (2021). Chapter-6 Chemistry and recent advances in management of acid sulphate soils. Latest Trends in, 95.

Fageria, N. K., & Baligar, V. C. (2003). Fertility management of tropical acid soil for sustainable crop production. In Handbook of soil acidity (pp. 373-400). CRC Press.

Fullen, M. A., & Catt, J. A. (2014). Soil management: problems and solutions. Routledge.

Ganeshamurthy, A. N., Kalaivanan, D., Satisha, G. C., & Peter, K. V. (2016). Management of fruit crops in acid soils of India. Innovations in horticultural sciences. New India Publishing Agency, New Delhi, India, 539-558.

Gitari, H. I. (2013). Lime and manure application to acid soils and their effects on bio-chemical Soil properties and maize performance at Kavutiri-Embu County (Doctoral dissertation, MSc Thesis, Kenyatta University, Kenya).

Harter, R. D. (2007). Acid soils of the tropics. ECHO Technical Note, ECHO, 11.

Haynes, R. J. (1982). Effects of liming on phosphate availability in acid soils: a critical review. Plant and soil, 68, 289-308.

Helyar, K. R. (1991). The management of acid soils. In Plant-Soil Interactions at Low pH: Proceedings of the Second International Symposium on Plant-Soil Interactions at Low pH, 24–29 June 1990, Beckley West Virginia, USA (pp. 365-382). Springer Netherlands.

Kochian, L. V., Piñeros, M. A., Liu, J., & Magalhaes, J. V. (2015). Plant adaptation to acid soils: the molecular basis for crop aluminum resistance. Annual Review of Plant Biology, 66, 571-598.

Merry, R. H., & Sabljic, A. (2009). Acidity and alkalinity of soils. Environmental and ecological chemistry, 2, 115-131.

Myers, R. J. K., & De Pauw, E. (1995). Strategies for the management of soil acidity. In Plant-Soil Interactions at Low pH: Principles and Management: Proceedings of the Third International Symposium on Plant-Soil Interactions at Low pH, Brisbane, Queensland, Australia, 12–16 September 1993 (pp. 729-741). Springer Netherlands.

Osman, K. T., & Osman, K. T. (2018). Acid soils and acid sulfate soils. Management of soil problems, 299-332.

Sarangi, S. K., Mainuddin, M., & Maji, B. (2022). Problems, management, and prospects of acid sulphate soils in the Ganges Delta. Soil Systems, 6(4), 95.

Shaaban, M. (2024). Acidic Soils. In Planet Earth: Scientific Proposals to Solve Urgent Issues (pp. 293-306). Cham: Springer International Publishing.

Sharma, U. C. (2013). Soil Acidity (Vol. 7). Springer Nature.

Sharma, U. C., Datta, M., & Sharma, V. (2025). Global Status and Extent of Acid Soils. In Soil Acidity: Management Options for Higher Crop Productivity (pp. 49-119). Cham: Springer Nature Switzerland.

Somani, L. L. (1996). Crop production in acid soils (No. Ed. 1). Agrotech Publishing Academy.

Sumner, M. E., Fey, M. V., & Noble, A. D. (1991). Nutrient status and toxicity problems in acid soils. Soil acidity, 149-182.

Von Uexküll, H. R., & Mutert, E. (1995). Global extent, development and economic impact of acid soils. Plant and soil, 171, 1-15.

Zade, S., Gourkhede, P. H., Vaidya, P. H., Singh, R. S., Sinha, S. K., Kumar, A., & Kumar, V. (2021). Problem Soils And Their Management Practices.

5

Acid Sulphate Soils

Introduction

Acid sulphate soils are formed when pyrite within a soil layer is oxidised, generating sulphuric acid. The oxidation of pyrite often results in yellow mottles of jarosite. The pH levels in greatly affected areas are often less than 4.0, and the associated environmental impacts include fish kills, retarded growth of crops and changes in water chemistry.

Acid soils are distributed in tropical and subtropical regions, constituting approximately 30% of the ice-free land in the world, which is 50% of the world's arable land. In China and India, 212 Mha or 12% of the agricultural land is classified as acidic. In India, about 28% of the total geographical area is affected by acidity of different degrees. Acid sulphate soils (ASSs) are a group of these soils having high soil acidity and other soil function limitations. These soils were termed as Kattecleigrondon or Kattakali by the Dutch farmers in the seventeenth century meaning 'Cat Clays' in English for soils of some reclaimed areas that became gradually highly acidic and developed prominent yellowish mottles and crusts composed of jarosite and related sulphates; in northern Germany, similar clays were called Maibolt.

Occurrence

They occur in many regions of the world, mostly along coastal areas where the land is inundated by sea water. Countries having extensive area of acid sulphate soils are:

Vietnam	-	2.0 m ha
Indonesia	-	2.0 - 3.0 m ha
Brazil	-	24.0 m ha
Thailand and India	-	0.81 m ha
Malesia	-	0.20 m ha

Generally acid sulphate soils are found in coastal areas where the land is inundated by salt water. In India, acid sulphate soil is, mostly found in Kerala, Orissa, Andhra Pradesh, Tamilnadu and West Bengal.

Murthy (1971) reported three types of these acid sulphate soils namely

i. **Kari soils:** Occupying about 80,000 ha mostly in **Kuttanad** (the rice bowl of Kerala).

ii. **Pokhali saline acid sulphate soils**: Occupying about 26,400 ha and

iii. **Swamp acid sulphate soils:** Occupying about 2,500 ha is found in Kerala state.

Table 1: International classification of Acid Sulphate Soils

Climate zone	Soil Taxonomy	Main characteristics
Potential acid sulphate soils		
World Wide	Typic Sulfaquents	pH (1:1 water) of dried soil < 3.5 within 50 cm if n > 1.0 within 30 cm if n < 0.7
World Wide	Sulfic Hydraquents	pH (1:1 water) of dried soil < 4.5 in upper 25 cm, or more acid between 50 and 100 cm; in n > 0.7 between 20 and 50 cm
Wet tropics & monsoon	Sulfic Trapaquents	Acidity requirements of sulfic hydraquents, but n < 0.7 between 20 and 50 cm
Temperate	Sulfic Fluvaquents	Same as Sulfic Trapaquents
World wide	Typic Sulfihemists	Sulfidic materials within 100 cm; pH (1:1 water) of dried soil <3.5

Mineralogy of acid sulphate soils

Studies conducted by Ghosh *et al.,* (1976) on Kari soils of Kerala reveal that Kaolinite is the dominant clay mineral (34.3%) associated with Smectite (18.32%). Besides these Illite (6- 12%), Chlorite (0-11%), Vermiculite (0-5%), Amphibole (0-4%), Gibbsite (0-17%), Quartz (0- 2%) and Feldspar (0-2%) were also present.

Types of acid sulphate soils

a) **Potential acid sulphate soils:** Potential acid sulphate soils which have not been oxidised by exposure to air are known as potential acid sulphate soils (PASS). They are neutral in pH (6.5–7.5), contain unoxidised iron sulphides, are usually soft, sticky and saturated with water and are usually gel-like muds but, can include wet sands and gravels have the potential to produce acid if exposed to oxygen.

b) **Actual acid sulphate soils:** When PASS is exposed to oxygen, the iron sulphides are oxidised to produce sulfuric acid and the soil becomes strongly acidic (usually below pH 4). These soils are then called actual acid sulphate soils (AASS). They have a pH of less than 4, contain oxidised iron sulphides, vary in texture and often contain jarosite (a yellow mottle produced as a by-product of the oxidation process).

Formation of Acid Sulphate Soils

Land inundated with waters that contain sulphates, particularly salt waters, accumulate sulphur compounds, which in poorly aerated soils are bacterially reduced to sulphates. Such soils are not usually very acidic when first drained in water.

When the soil is drained and then aerated, the sulphide (S_2^-) is oxidized to sulphate (SO_2^{-4}) by a combination of chemical and bacterial actions, forming sulphuric acid (H_2SO_4). The magnitude of acid development depends on the amount of sulphide present in the soil and the conditions and time of oxidation. If iron pyrite (FeS_2) is present, the oxidized iron accentuates the acidity but not as much as aluminium in normal acid soils because the iron oxides are less soluble than aluminium oxides and so hydrolyse less.

The oxidation of mineral sulphides (pyrites – FeS_2) in soil is slow and non-bacterial until the soil pH reaches 4.0. But below pH 4.0, the bacteria *Thiobacillus ferroxidans* are the most active oxidizers and the acidity builds up rapidly. The chemical reactions are as follows

Non-Biological (chemical)

$2FeS_2 + 2H_2O + 7O_2$ --------------→ $2\ FeSO_4 + 2\ H_2SO_4$

(Pyrites) (Ferrous sulphate)

Biological (accelerated by bacteria)

Thiobacilus

Ferrooxidans

$4FeSO_4 + O_2 + 2\ H_2SO_4$ -----------→ $2Fe_2(SO_4)_3 + 2H_2O$

(Ferric sulphate)

Non-biological (Rapid in acid pH)

$FeS_2 + 7Fe_2(SO_4)_3 + 8H_2O$ -------→ $15FeSO_4 + 8H_2SO_4$

(Ferrous sulphate) (Sulfuric acid)

The extent of acid development depends on the amount of sulphide in the soil and the conditions and time of oxidation.

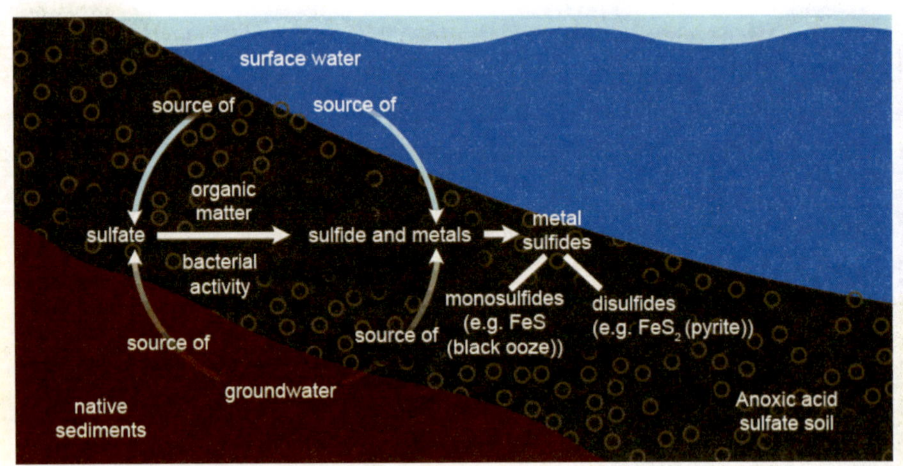

Properties of Acid Sulphate Soil

Acid sulphate soils contain a sulphuric horizon, which has a pH of 3.5 in 1:1 soil to water extract and a high sulphide content (yellow colour mineralogy). Mottle or coating of jarosite and ochre in soils.

Poblems of Acid Sulphate Soils and Harmful Effects on Plant Growth

Acid sulphate soils pose chemical, physical and biological problems to crops.

A. The direct effects of severe acidity due to draining and aeration of these soils are

1. Increased solubility and toxicity of A1, Fe, Mn and possibly also H ions.
2. Decreased availability of phosphorus due to formation of iron and aluminium phosphates.
3. Low base status and nutrient deficiencies
4. Salinity.

B. Under flooded conditions for e.g. under rice cultivation or fish ponds, acidity is reduced but new problems arise, they are:

1. Iron Toxicity.
2. Hydrogen sulphide toxicity. H_2S gas often causes a rice disease called **akiochi** or **brusone,** which prevents absorption of nutrients by rice plant roots.
3. CO_2 and Organic acid toxicity.

C. Physical problems arise mainly through

1. Inhibition of root development in acid sulphate horizon.

2. Crop suffers from stress, soil ripening is arrested, clay and organic soils remain soft, unable to bear heavy loads. They are poorly structured and therefore these soils are poorly drained.
3. Field drains may be blocked by iron oxide (ochre) deposits.

D. Unsuitable conditions for most micro-organisms, impede the release of nutrients from soil organic matter, crops under stress are especially susceptible to diseases.

E. Crops suffer from severe physiological stress.

1. Plant roots are unable to enter the acid layer hence water stress similar to the effects of salinity are observed.
2. Plant growth is reduced, leaves dieback and scorch.
3. Particularly root system is stunted or sometimes deformed.
4. In rice the tillers of young plants may die, bronzing of the leaf tips often occurs.

Causes of low productivity in acid soils

- Low productivity of these soils is due to :
- Injury of hydrogen ions.
- Low pH causing
- Impaired availability and absorption of plant nutrients.
- Increased solubility of iron, aluminum and manganese resulting in toxicity.
- Decreased availability of phosphorus and molybdenum.
- Low base saturation due to leaching.
- Abnormal biotic factors causing impaired mycorrhiza and virulence of plant diseases.
- Salt injury.
- Telicity due to hydrogen sulphide (H_2S).
- Production of organic acids.

Plant roots are directly injured due to high concentration of hydrogen ions particularly if the pH is < 4.0. But plants can tolerate concentration of hydrogen ions (H^+) if concentration of polyvalent cations is less. It causes suppressed root growth and root branching because of adverse effect on cell cytoplasm and cell permeability. Numbers of plant roots are less, thick and dull grey to brown in colour. Nutrient absorption capacity of root is reduced. If the pH is around 4.0 rubber, oil palm, coconut, cassava and banana can be grown. Paddy can also be grown under submergence as, it can tolerate high concentration of aluminum (1.2 to 25 mg kg^{-1}) if supply of other nutrients can be maintained because concurrent increase in pH due to continuous submergence lowers the concentration of aluminum. The concentration of aluminum which was 74.5 mg kg^{-1} at pH 3.11 is reduced to 0.3 mg kg^{-1} at pH 4.0. This may be because ferric hydro oxide is reduced to ferrous hydro oxide and the OH released causes the precipitation of aluminum.

$Fe(OH)_3 \rightarrow Fe(OH)_2 + OH^-$

$Al^{3+} + OH^- \rightarrow Al(OH)^{++}$, $Al(OH)_2^+$ and $Al(OH)_3$

- Paddy can also tolerate high concentration of Fe^{2+} (200 mg kg^{-1}). But excess of H_2S present in this range causes Akiochi disease as a result of which respiratory activity of root is reduced, plants become deficient in K, N, Si and bases, iron content in the plant increases to toxic levels and plants are infected with Helminthosporium which causes wide spread damage. H_2S can also cause suffocation disease which can be corrected by application of nitrate nitrogen. Another disease called Brusone also develops which is caused by *Piricularia oryzae* and the plant suffers.
- High concentration of aluminum makes the plant roots brittle, devoid of branching with thick root tips and brown lateral roots. Aluminum is attached to phosphate in the DNA structure which inhibits cell division and protein synthesis. It also causes precipitation of nucleic acids and increases the water potential of plants.
- High concentration of manganese decreases translocation of calcium to leaves and increases oxidation of Indole Acetic Acid (IAA) and hence growth of plant tops is affected. Activity and functioning of Mn activated enzyme system is reduced. Excess of Mn causes marginal chlorosis of young leaves and necrotic spots on leaves. However, since considerable quantities of Mn is lost during formation of these soils (due to leaching) very few of these soils suffer from Mn toxicity.
- Organic acids, such as acetic acid, n-butyric acid and propionic acid accumulate in these soils. These acids are phytotoxic.
- The concentration of soluble salts also increases particularly when pyritic soils undergo oxidation and drying. The electrical conductivity (EC) may be as high as 10 dSm^{-1}. Periodic flooding of acid sulphate soils with sea water may further accelerate the salt problem. Because of these reasons the productivity is low and :

  1. Lower yield of crops
  2. Lower chances of increasing yields due to fertilizer application

3. Further complications arising out of dry spells accentuating the soil acidity.
4. Drains are blocked due to ochre.
5. Inundation due to sea water.
6. Continuous water logging causing H_2S toxicity.

Soil quality problems of acid sulphate soils

In general, acid sulphate soils are unproductive or low productivity, it may be due to one or more of the unfavourable factors i.e. soil acidity, salinity, aluminium toxicity, iron toxicity, low content of major nutrients, low base status, and hydrogen sulphide toxicity.

- **Soil Acidity:** The acid sulphate soils may be due to the direct effect of hydrogen (H+) ions, especially below pH 3.5 to 4. However, aluminum toxicity is probably more important in this pH range.
- **Soil Salinity:** Acid sulphate soils in tidal areas are often affected by salinity. Salinity shows the toxicity, by weakening the plants and increasing iron and aluminium concentration in solution.
- **Aluminium Toxicity:** One cause of stress on the growth of certain plant species is aluminium toxicity. A high Al level affects cell division, disrupts certain enzyme systems, and hampers uptake of phosphorus, calcium and potassium. Most plants grown on acid sulphate soils which have a pH below 4 suffer from Al toxicity.
- **Iron Toxicity:** Dissolved iron in excess of 300-400 mg kg^{-1} is toxic to rice crop.
- **Low Nutrient Contents:** In the absence of iron and aluminium toxicity and harmful salinity, phosphorus deficiency is the most important problem of acid sulphate soils. Supply of nitrogen increases the phosphate response.
- **Low Base Status:** During the formation of acid sulphate soils, bases are removed as sulphate and most of the exchange complex is occupied by aluminium. Therefore, acid sulphate soils are likely to be deficient in Ca and K.
- **Hydrogen Sulphide Toxicity:** Hydrogen sulphide has been shown to be toxic to the rice plant through its suppression of the oxidizing power of the roots.

Management and Reclamation of Acid Sulphate Soils

Management techniques are extremely variable and depend on the extent of acid formation, the thickness of sulphide layer, leaching possibilities and value of the land area. The general reclamation measures are as follows.

1. **Avoidance:** Avoidance is the most preferred management strategy for management of acid sulphate soil, and should be considered at all sites. Acid sulfate soils are inert when left in waterlogged, undisturbed conditions. Avoidance is often the most environmentally responsible and cheapest option.
2. **Minimisation of Disturbance:** If acid sulphate soils cannot be avoided for crop production, their disturbance should be minimised. Completion of a detailed acid sulphate soil investigation is essential for minimisation of disturbance to be effective. Once the site has been adequately characterised, strategies that minimise the disturbance can be investigated.
3. **Neutralisation:** Liming is the primary and most important way to reclaim acid sulphate soil. It involves the physical incorporation of neutralising/alkaline materials into the soil. Lime has an alkaline pH and buffers any acid produced whilst raising the soil pH to acceptable levels. Proper mixing the appropriate amount and type of lime into disturbed acid sulphate soils will neutralise soil acidity produced. If acid sulphate soils are leached during early stage of acidification, lime requirement are lowered.
4. **Re-flooding:** The objective of re-flooding is to neutralise actual acidity and reduce the pyrite oxidation rate. Re-flooding relies on establishing conditions where the reduction of the Fe, Mn, S and N can take place. The reduction of these elements is responsible for the increase in pH commonly observed in acid soils after water logging. Re-flooding can also be used as a water table height-management tool to prevent the oxidation of PASS or further oxidation of ASS.
5. **Seawater re-flooding:** Re-flooding with seawater may trap the existing acid leachate and force it deeper into the soil profile, limiting the export of oxidation products and slow diffusion into the tidally exchanged seawater. The advantage of this approach is that it is simple, cheap and improves the situation even if further improvement is eventually needed.
6. **Hydraulic separation:** Hydraulic separation is suitable for sandy material containing iron sulphides. Sluicing or hydrocycloning are used to hydraulically separate the sulphides from the sandy materials.

This technique is very much effective in areas where the sediments contain <10–20% clay and silt, and have low organic matter content. The separated sulfidic material extracted via the process requires special management involving either neutralisation or strategic reburial.

7. **Bioremediation:** By re-establishing reducing conditions within the bunded area, pyrite oxidation may be reversed by sulfate-reducing bacteria. In effect, it would re-establish the sulphide formation processes that operate in the mangrove soils outside the bund wall. Bioremediation causes chemical changes in the water and soil, and in sediment that may accumulate. Bioremediation would be a natural process and cost-effective if *in-situ* microbial generation of acid-neutralising capacity is significant.
8. **Cover in-situ soils with clean fill:** If groundwater levels are not affected by earthworks, undisturbed *in-situ* potential acid sulphate soil can be covered with a significant volume of clean fill. A minimum depth of fill cannot be specified for residential or commercial/industrial development. A suitable depth of fill should rather be determined on a site specific basis, dependent on the severity and extent of acid sulphate soil, as identified in the investigation. Once a site has been covered by clean fill, any associated infrastructure may be placed within the fill, thereby not disturbing any *in-situ* acid sulphate soil by excavation or dewatering.
9. **Flooding and intermittent drainage:** Soils may be flooded (anaerobic) or buried in water to maintain a saturated state to minimize acid sulphate soil. This solution almost limits the use of the area to rice growing. But, sometimes drought occurs unfortunately and causes soil acidification in short time period. The water used to flood the potential acid sulphate soils often develop acidity and injure crops. Rice grown under intermittent drainage had healthier root systems, less empty grains, heavier weight per panicle. Rice in continuous submergence showed strong bronzing symptoms but had more tillers.
10. **Water table management:** Sometimes in acid sulphate soil, non-acidifying layer covers sulphuric horizon. Then drainage to keep only the sulphuric layer under water (anaerobic) is possible. By raising the water table, after damage has been inflicted due to over-intensive drainage, the soils can be restored.
11. **Deep soil mixing:** It is carried out with a large diameter (one to three metres) hollow-flight auger It also has special mixing '*paddles*' which mixes soil. As holes are drilled into the soft substrate lime or cement and

a variety of binding agents are mixed with the soil slurry which form solid supportive columns in the soil after cementation.

12. **Growing of suitable crops:** Rice is the most preferable crop which is highly acid tolerant. Adoption of rice crop in acid sulphate soils increases the pH of soil and thus reduces the iron and aluminium toxicity. Acid sulphate soils with a widely spaced subsurface drainage system have yielded promising results for the cultivation of upland rice, peanut and soybean.

Conclusion

Acid sulphate soils pose significant environmental and agricultural challenges due to their high acidity, metal toxicity, and potential to degrade ecosystems. Effective management strategies, including proper drainage, liming, and sustainable land-use practices, are essential to mitigate their adverse effects. By implementing science-based solutions and monitoring soil conditions, the harmful impacts of these soils can be minimized, ensuring healthier ecosystems and improved agricultural productivity. Addressing acid sulphate soils requires a proactive approach to balance land use with environmental preservation for long-term sustainability.

References

Das, S. K., & Das, S. K. (2015). Acid sulphate soil: management strategy for soil health and productivity. Popular kheti, 3(2), 2-7.

Dash, A. K., Singh, C., & Varma, S. (2021). Chapter-6 Chemistry and recent advances in management of acid sulphate soils. Latest Trends in, 95.

Dent, D. L. (1986). Acid sulphate soils: a baseline for research and development (No. 39). ILRI.

Dent, D. L., & Pons, L. J. (1995). A world perspective on acid sulphate soils. Geoderma, 67(3-4), 263-276.

Fageria, N. K., & Baligar, V. C. (2003). Fertility management of tropical acid soil for sustainable crop production. In Handbook of soil acidity (pp. 373-400). CRC Press.

Fullen, M. A., & Catt, J. A. (2014). Soil management: problems and solutions. Routledge.

Ganeshamurthy, A. N., Kalaivanan, D., Satisha, G. C., & Peter, K. V. (2016). Management of fruit crops in acid soils of India. Innovations in horticultural sciences. New India Publishing Agency, New Delhi, India, 539-558.

Gitari, H. I. (2013). Lime and manure application to acid soils and their effects on bio-chemical Soil properties and maize performance at Kavutiri-Embu County (Doctoral dissertation, MSc Thesis, Kenyatta University, Kenya).

Grealish, G. J., & Fitzpatrick, R. W. (2013). Acid sulphate soil characterization in N egara B runei D arussalam: a case study to inform management decisions. Soil Use and Management, 29(3), 432-444.

Indira, B. V. N., & Covilakom, M. T. K. (2013). Characterization of acidity in acid sulphate soils of Kerala. Journal of life sciences, 7(8), 907.

Mayakaduwage, S., Vithana, C. L., Mosley, L. M., & Vithanage, M. (2023). Management of Acid and Acid Sulfate Soils. In Soil Constraints and Productivity (pp. 381-407). CRC Press.

Merry, R. H., & Sabljic, A. (2009). Acidity and alkalinity of soils. Environmental and ecological chemistry, 2, 115-131.

Michael, P. S., Fitzpatrick, R. W., & Reid, R. J. (2016). The importance of soil carbon and nitrogen for amelioration of acid sulphate soils. Soil Use and Management, 32(1), 97-105.

Myers, R. J. K., & De Pauw, E. (1995). Strategies for the management of soil acidity. In Plant-Soil Interactions at Low pH: Principles and Management: Proceedings of the Third International Symposium on Plant-Soil Interactions at Low pH, Brisbane, Queensland, Australia, 12–16 September 1993 (pp. 729-741). Springer Netherlands.

Osman, K. T., & Osman, K. T. (2018). Acid soils and acid sulfate soils. Management of soil problems, 299-332.

Panhwar, Q. A., Naher, U. A., Shamshuddin, J., Radziah, O., & Hakeem, K. R. (2016). Management of acid sulfate soils for sustainable rice cultivation in Malaysia. Soil science: Agricultural and environmental prospectives, 91-104.

Prasittikhet, J., & Gambrell, R. P. (1990). Acidic sulfate soils. In Acidic Precipitation: Soils, Aquatic Processes, and Lake Acidification (pp. 35-62). New York, NY: Springer New York.

Sarangi, S. K., Mainuddin, M., & Maji, B. (2022). Problems, management, and prospects of acid sulphate soils in the Ganges Delta. Soil Systems, 6(4), 95.

Shaaban, M. (2024). Acidic Soils. In Planet Earth: Scientific Proposals to Solve Urgent Issues (pp. 293-306). Cham: Springer International Publishing.

Sharma, U. C. (2013). Soil Acidity (Vol. 7). Springer Nature.

Sharma, U. C., Datta, M., & Sharma, V. (2025). Global Status and Extent of Acid Soils. In Soil Acidity: Management Options for Higher Crop Productivity (pp. 49-119). Cham: Springer Nature Switzerland.

Thakur, R. Acid Sulphate Soils and Their Management.

Zade, S., Gourkhede, P. H., Vaidya, P. H., Singh, R. S., Sinha, S. K., Kumar, A., & Kumar, V. (2021). Problem Soils And Their Management Practices.

6

Calcareous Soils and Their Management

Introduction

The word "calcareous" has been commonly used as the generic description of all materials containing calcium carbonates greater than 10%. In the context of agricultural problem soils, calcareous soils are soils in which a high amount of calcium carbonate dominates the problems related to agricultural land use. They are characterized by the presence of calcium carbonate in the parent material and by a calcic horizon, a layer of secondary accumulation of carbonates (usually Ca or Mg) in excess of 15% calcium carbonate equivalent and at least 5% more carbonate than an underlying layer. Calcareous soils are common in the arid areas of the earth (FAO, 2016) occupying >30% of the earth's surface. In the world Reference Base (WRB) soil classification system calcareous soils may mainly occur in the Reference soil group of calcisols.

Formation / genesis of calcareous soils

The soil are formed largely by the weathering of calcareous rocks and fossil shell beds like varieties of chalk, marl , lime stone and frequently a large amount of phosphates. Soils also can become calcareous through long term irrigation with water contains small amounts of dissolved $CaCO_3$ that can accumulate with time.

The secondary $CaCO_3$ are formed under arid and semi-arid climatic conditions, when the carbonate concentration in the soil solution remains High Accumulation starts in the fine and medium sized pores at the surface of contact between the soil particles. This accumulation may be rather concentrated in a narrow zone of the solum or more dispersed depending upon the quantity and frequency of rainfall, topography, soil texture and vegetation. In high rainfall area i.e. in highly weathered soils, $CaCO_3$ is dispersed from the surface horizons. While in arid and semi-arid regions due to low rainfall and limited leaching carbonates may be present in relatively high amounts. In medium deep and deep black clay soils carbonates are generally accumulated in subsurface horizons. However in some eroded shallow black soils they may be exposed right at the surface of soils. In some soils $CaCO_3$ deposits are concentrated into layers that may be very hard and impermeable to water also called caliche. These caliche layers are formed by rainfall, leaching the salts to particular depth in the soil at which the moisture content is so low that the carbonates precipitate. They are also formed by salts moving upward from water table (caused by irrigation) and precipitating at the top of capillary tinge.

Distribution of calcareous soil

The total extent of calcisols is estimated at 800 million hectares worldwide, mainly concentrated in arid or Mediterranean climates. However, the total area of calcareous soils is difficult to estimate because many calcisols occur together with solon chalks that are actually salinized calcisols or Solonetz (sodic soil) and/or with other soils of carbonate enrichment that do not key out as calcisols.

The primary sources of calcareous soil are reef-building corals, coral algae, and bone fragments of other marine organisms deposited in situ or transported from nearby sources. They are mainly distributed in the north and south sides of the equator in the Pacific Ocean, the Indian Ocean, and the Atlantic Ocean. These soils are generally located between 30^0S and 30^0N along tropical and subtropical marine regions, island coasts, and sea areas. The total area of coral reefs in the world is about 284,300 km^2, of which 91.9 % are located in the Indian Ocean and Pacific Ocean (including the Red Sea, the Indian Ocean, Southeast Asia, and the Pacific Ocean). Calcareous soil surveys in different sea areas worldwide are mainly concentrated in the Persian Gulf, South China Sea, Western Australia, and Puerto Rico

Characteristics

- Usually have alkaline soil reaction (PH >7.0) & High buffering capacity.
- Soils are dominated by carbonates of calcium and magnesium mainly soil contain $CaCO_3$ in free form $CaCO_3$ may occur indifferent forms (powder/nodules) with reduced availability of N, P, K, S, Fe, Zn, and B.
- The phosphorus losses are rapid during initial stages but the intensity of P fixation decreased with the increase in time. The higher soluble phosphate fertilizers are more porn to their fixation because they are more active to chemical reaction with the constituent of the soil.
- Iron deficiency due to high $CaCO_3$ leads to chlorosis also called lime induced iron chlorosis and reduced rate of seed germination.
- Decreased water holding capacity (WHC) due to alteration in soil structure, formation of hard pan.
- Flocculation due to enough Ca and Mg present in calcareous soil increases aggregate stability and when a large percentage (>30%) of $CaCO_3$ is present in the clay fraction, the soils WHC can be reduced.

- Surface crusting and sub-surface hard pan formed restrict aeration, infiltration and affects rate of seedling emergence and establishment.
- Activity of rhizosphere micro-organisms is reduced under less moisture conditions.

Physical properties of calcareous soil

The basic physical properties of calcareous soils, such as specific gravity (GS), porosity ratio (e), particle morphology, and cementation characteristics, are also different compared to other soils due to the global differences of sedimentary environment and material composition.

Specific gravity

Soil particle specific gravity (G_S) is determined by the mineral composition of the soil, and G_S is mainly used to calculate the void ratio and compressibility. Calcareous soil particles contain internal pores, and when the particles are broken, the compressibility of the calcareous soil foundation will change, which will then affect the engineering performance. It is suggested that the influence of the GS of a calcareous soil on the design of reef foundation and revetment slope should be fully considered during the construction work on islands and reefs.

Void ratio

A greater void ratio in calcareous soil results in greater compressibility and deformation, indicating more obvious particle crushing characteristics.

Particle morphology

Calcareous soil particle morphology has significant fractal characteristics, which are positively correlated with particle size. Its fractal dimension ranges from 0.95 to 1.07. The shape factors (aspect ratio, sphericity, and concavity) of calcareous soil particles appear to increase with increasing pressure.

Production constraints

Crust formation

Crust formed on the soil's surface is one of the most important component of the soil. This is true in the case of arid, semiarid and the moderate environment as well (Rahmonov, 2006). Gillette proved with his experiments carried out in wind tunnel that surface covered with crust is less vulnerable for wind erosion (Gillette *et al.*, 1982). There are two main types of surface crusts: one of them is formed out in natural way and the other one is formed out by anthropogenic effects. The former one can be divided into two other groups: the one that is formed by moisture (this is called as physical) and the one that has a biological origin (Goossens, 2004). The latter one forms out as a result of the cementing effect of different algae, lichens and bacteria living in soil. These microplants can bind soil particles and don't let them start to move and protect them from the bombarding effect of other grains. Excretion of organic matter- for example polysaccharides- can increase the strength of such type of crust.

The biological crusts are important in the arid and semiarid regions, because the vegetation cover is not developed enough to provide the stability of surface. Investigations carried out in desert regions established that the biological crust has more strength and provide better protection than the other one evolved by the effect of precipitation. Crusting which takes place at the soil surface hinders seedling rate of emergence and percentage.

Ammonia Volatilization

In the soil solution, the ammonia gas is in equilibrium with ammonium ions according to the following reversible reaction:

$$\mathbf{NH_4^+ + OH^- \rightarrow H_2O + NH_3}$$

From the above equation it is obvious that ammonia volatilization will be more pronounced at high pH levels (pH>7.5). On the other hand, ammonia-gas-producing amendments will drive the reaction to the left, raising the pH of the solution in which they are dissolved. When a moist soil dries, the water is moved from the right-hand side of the equation, again driving the equation to the right, encouraging ammonia volatilization. Compared with NH_3 volatilization loss from N fertilizers, NH_3 loss from N mineralized from organic N is relatively small. Urea hydrolysis can increase soil pH thus encourage NH_3 volatilization. Therefore, in neutral and acidic soils, NH_4^+-N containing fertilizers are less subject to volatilization loss than urea-containing fertilizers. However, in calcareous soils, soil solution pH is buffered at about 7.5, a condition favour volatilization loss for even NH_4^+ containing fertilizers, for example:

$$\mathbf{(NH_4)_2SO_4 + CaCO_3 \rightarrow 2NH_3 + H_2O + CO_2 + CaSO_4}$$

Since $CaSO_4$ is only slightly soluble, the reaction proceeds to the right and NH_3 volatilization is favoured. Similar reactions occur with other NH_4^+ containing fertilizers that produce insoluble Ca precipitates (e.g., $[NH_4]_2HPO_4$). In comparison, volatilization losses are not as great if the NH_4^+ containing fertilizers (e.g., NH_4NO_3, NH_4Cl) produce soluble Ca reaction products. Ammonia volatilization losses also increase with increasing fertilizer rate and with liquid compared with dry N sources. Volatilization of NH_3 is much greater with broadcast applications compared with subsurface or surface band methods. Immediate incorporation of broadcast N greatly reduces the NH_3 volatilization potential.

Soil colloids, both clay and humus, adsorb ammonia gas, so ammonia losses are greatest where low quantities of these colloids are present or where the ammonia is not in close contact with the soil. For these reasons, ammonia loss can be quite large from sandy soils and from alkaline or calcareous soils, especially when the ammonia-producing materials are left at or near the soil surface and when the soil is drying out. High temperatures, as often occur on the surface of the soil, also favors the volatilization of ammonia. Incorporation of manure and fertilizers into the top few centimeters of soil can reduce ammonia losses by 25-75% from those that occur when the material are left on the soil surface.

Precipitation of Soluble Phosphate

Phosphorus (P) is an essential macronutrient, being required by plants in relatively large quantities (~0.2 to 0.8%). Potassium and nitrogen are the only mineral nutrients required in larger quantities than P. providing adequate P to plants can be difficult, especially in calcareous soil. Alkaline soil is defined as soil with pH greater than neutral, typically 7.5 to 8.5. Calcareous soil is defined as having the presence of significant quantities of free excess lime (calcium or magnesium carbonate). Lime dissolves in neutral to acid pH soil, but does not readily dissolve in alkaline soil and, instead, serves as a sink for surface adsorbed calcium phosphate precipitation. The bioavailability of P is strongly tied to soil pH. The formation of iron and aluminum phosphate minerals results in the reduced solubility of P in strongly acidic soil, improving as pH approaches nearly neutral. This maximum solubility and plant availability of P at pH 6.5 declines again as the pH increases into the alkaline range. This effect of reduced P availability in alkaline soil is driven by the reaction of P with calcium, with the lowest solubility of these calcium phosphate minerals at about pH 8. The presence of lime in alkaline soil further exacerbates the P availability problem. The lime in calcareous soil reacts with soil solution P to form a strong calcium phosphate bond at the surface of the lime. These calcareous soils are common in arid and semi-arid regions with little rainfall. The resulting effect of low P solubility in calcareous soil is relatively poor fertilizer P efficiency. Plants grown in these conditions can be stunted with shortened internodes and poor root systems due to P deficiency. Deficiency symptoms are sometimes observed as a darkening of the leaf tissue, although it is more common to observe yield loss with no readily seen symptom. Simply adding fertilizer Pat "normal" rates and with conventional methods may not result in optimal yield and crop quality. Several fertilizer P management strategies have been found to improve P nutrition for plants grown in alkaline and calcareous soil.

Precipitation of Iron Compounds

Iron (Fe) is one of the most studied elements in mineral nutrition of plants. Although it's relatively high abundance in the earth's cultivated soil, plant iron acquisition is often impaired, a fact resulting in severe crop losses. According to Lindsay (1995) the total Fe in soils is clearly higher than the soluble Fe required for optimal growth, which is at approximately 10-8 M in the soil solution. However, he added that the availability of free Fe (III) in soils is basically low: it depends largely on pH and on redox potential and increases with low soil pH and low redox potential. In calcareous soils, the concentration of free Fe (III) is extremely low (about 10-10M). Plants grown in calcareous soils suffer from Fe deficiency chlorosis worldwide, frequently showing yellow leaves and older green leaves. Apart from leaf chlorosis, Fe deficient plant may show depressed leaf formation and growth even when the youngest leaves are green. The frequent finding of high Fe concentrations in chlorotic leaves as well as fairly high root Fe concentrations is puzzling. Both total and active soil carbonates, as a dominant active solid phase in calcareous soil, have a significant effect on iron availability.

Management of Calcareous soil

1. Since the calcareous soils develop in regions of low rainfall, they must be irrigated to be productive. Potential productivity of calcareous soils is high where adequate water and nutrients can be supplied. The management of water and nutrients is the main production challenge. Optimum amounts of water for plant growth have to be provided without wastage, and salts which may affect plant growth have to be controlled.
2. Chiselling to break the hard cemented layer of $CaCO_3$ in calcareous soils increases permeability, helps to reclaim these soils.
3. Application of organic manures such as FYM, compost and green manuring.
4. Apply acid or acid forming amendments like sulphur, pyrites, $FeSO_4$, $Al_2(SO_4)_3$ to reclaim the soils effectively by dissolving $CaCO_3$
5. Use of acidic or acid forming fertilizers
6. Foliar application of micronutrients viz., Zn and Fe
7. Apply adequate amount of N, P, K and Mg fertilizers, split application of phosphates.
8. For effective reclamation through amendments and to improve potential productivity of crop in these soils, provide optimum irrigation.
9. Application of press mud compost 5 t/ha once in three years before ploughing.

10. Soil analysis/testing for $CaCO_3$ content in different layers of soil profile is necessary before planning of horticultural crops.

Nutrient management in calcareous soil

Nitrogen

1. Availability of plant nutrients is generally found decreased in calcareous soil due to its alkaline nature.
2. Most of plant nutrients are available when soil pH ranges between 6.5 to 7.5 under high pH of availability N to plant decreases due to reduced rate of nitrification and loss of N through denitrification process. In soil ammonia converts first to nitrate and then to nitrate and becomes available to plant.
3. Ammonium N fertilizers like ammonium sulfate and ammonium phosphate are useful when pH of soil is less than 7.5 as during nitrification process H^+ ions are released which neutralize the $CaCO_3$ and helps to reduce soil pH. But when these ammonical fertilizers are used in calcareous soil, nitrogen is lossed in the form of NH_3 as ammonical compounds turn into ammonia after reacting with $CaCO_3$ in soil.
4. Hence, use of ammonium sulphate, ammonium phosphate should be avoided in calcarcous soil. Instead of these ammonium nitrate and ammonium chloride are found useful as the loss of N is less when these sources of N are utilized. After knowing initial chloride status of soil, ammonium chloride should be used.

Overall cares need to be taken for management of N fertilizers in calcarcous soil

1. To avoid loss of N in the form of NH_3, added fertilizers should get mixed well within soil after ensuring proper moisture, fertilizers should be added. If moisture is less, then soon after fertilizers application supplemental irrigation is need to be given.
2. To avoid loss of N in NH3, urea should be added along with MOP or triple super phosphate. Use of sulphur coated or neem coated urea also found beneficial and it improves efficiency of N fertilizers.

Phosphorus

1. In general efficiency of phosphorus nutrient ranges between 15 -20 % and in calcareous soil its efficiency and availability is found very low. At pH 6 to 7.5 phosphorus is usually available due to higher pH, availability of P is reduced in calcareous soil and P often turns into tri calcium phosphate, magnesium phosphates which are less soluble in water.

2. As these insoluble compounds are formed after addition of P fertilizers in calcareous soil, its availability is decreased this is called as P fixation. These insoluble compounds are formed and retained within soil. As soil pH increases, rate of formation of these insoluble compounds increases and availability P decreases. Hence, to increase its availability P fertilizers are need to be added with organic matter and Use of PSB is also helpful to increase solubility of P in soil. Easily soluble sources like SSP, DAP should be used.
3. Band placement of P fertilizers near to roots and in granular form helps in increasing availability of P. Time of application of fertilizer is very important regarding plant growth. Plant must get Pat right time for development of roots. Addition of SSP along with FYM/compost to crops helps in increasing P availability and development of roots.

Potassium

Calcareous soil contains enough amount of potassium but due to higher concentration of calcium uptake of potassium ion is affected. Hence, deficiency of potassium is observed in plants. For example grapes become too acidic in calcareous soil due to less uptake of potassium. Therefore, potassium should be added in quantity more than its recommended dose under high calcium content in soil.

Micronutrients

The deficiencies of micronutrients are mostly saviour problems of calcareous of soil. Calcium carbonates can be easily fixes all the nutrients (Wahba *et al.* 2019). Thus, addition of micronutrients like, Zn, Cu, Fe, Mn and B would be helpful in increasing the yield. The deficiencies of micronutrients are normally corrected through soil or foliar application as described in table given below:

General recommendations of micronutrient fertilizers (Samal and Kumar 2020)

Micronutrients	Source of micronutrients fertilizers	
	Soil application	**Foliar application**
Zn	Zinc sulphate (25 kg ha^{-1})	0.5 % $ZnSO_4$ + 0.25 % lime
Fe	Iron sulphate (50 kg ha^{-1})	1.0 % $FeSO_4$ + 0.5 % lime
Cu	Copper sulphate (10 kg ha^{-1})	0.1 % $CuSO_4$ + 0.05 % lime
Mn	Manganese sulphate (10 kg ha^{-1})	1.0 % $MnSO_4$ + 0.25 % lime
B	Borax (10 kg ha^{-1})	0.2 % $ZnSO_4$

Conclusion

Calcareous soils present unique challenges for agriculture due to their high calcium carbonate content, which affects nutrient availability, soil structure, and plant growth. Effective management of these soils requires practices such as proper irrigation to prevent salinity, the use of acid-forming fertilizers to improve nutrient uptake, organic matter incorporation to enhance soil fertility, and the selection of tolerant crop varieties. By adopting these strategies, farmers can improve productivity and sustainability in calcareous soil environments, ensuring better crop yields and long-term soil health.

References

Ahmad, M., Ishaq, M., Shah, W. A., Adnan, M., Fahad, S., Saleem, M. H., ... & Hashem, M. (2022). Managing phosphorus availability from organic and inorganic sources for optimum wheat production in calcareous soils. Sustainability, 14(13), 7669.

Batarseh, M. (2017). Sustainable management of calcareous saline-sodic soil in arid environments: The leaching process in the Jordan Valley. Applied and Environmental Soil Science, 2017(1), 1092838.

Bolan, N., Srivastava, P., Rao, C. S., Satyanaraya, P. V., Anderson, G. C., Bolan, S., ... & Kirkham, M. B. (2023). Distribution, characteristics and management of calcareous soils. Advances in agronomy, 182, 81-130.

Gillette, D. A., & Passi, R. (1988). Modeling dust emission caused by wind erosion. Journal of Geophysical Research: Atmospheres, 93(D11), 14233-14242.

Goossens, D. (2004). Effect of soil crusting on the emission and transport of wind-eroded sediment: field measurements on loamy sandy soil. Geomorphology, 58(1-4), 145-160.

Hopkins, B., & Ellsworth, J. (2005, March). Phosphorus availability with alkaline/calcareous soil. In Western nutrient management conference (Vol. 6, No. 3-4, pp. 83-93). Idaho Falls, ID: University of Idaho.

Jalali, M., Merikhpour, H., Kaledhonkar, M. J., & Van Der Zee, S. E. A. T. M. (2008). Effects of wastewater irrigation on soil sodicity and nutrient leaching in calcareous soils. Agricultural water management, 95(2), 143-153.

Khan, M. A., Basir, A., Fahad, S., Adnan, M., Saleem, M. H., Iqbal, A., ... & Nawaz, T. (2022). Biochar optimizes wheat quality, yield, and nitrogen acquisition in low fertile calcareous soil treated with organic and mineral nitrogen fertilizers. Frontiers in Plant Science, 13.

Kosegarten, H., Wilson, G. H., & Esch, A. (1998). The effect of nitrate nutrition on iron chlorosis and leaf growth in sunflower (Helianthus annuus L.). European Journal of Agronomy, 8(3-4), 283-292.

Lindsay, W. L., & Ajwa, H. A. (1995). Use of MINTEQA2 for teaching soil chemistry. Chemical equilibrium and reaction models, 42, 219-239.

McCauley, A., Jones, C., & Jacobsen, J. (2005). Basic soil properties. Soil and water management module, 1(1), 1-12.

Obreza, T. A., Alva, A. K., & Calvert, D. V. (1993). Citrus fertilizer management on calcareous soils (p. 1993). Gainesville, FL, USA: Cooperative Extension Service, University of Florida, Institute of Food and Agricultural Sciences.

Renella, G., Adamo, P., Bianco, M. R., Landi, L., Violante, P., & Nannipieri, P. (2004). Availability and speciation of cadmium added to a calcareous soil under various managements. European Journal of Soil Science, 55(1), 123-133.

Sakin, E., & Yanardag, I. H. (2024). The Advantages and Disadvantages of Calcareous Soils. In" Agricultural Research Updates, 45, 147-162.

Tagliavini, M., & Rombola, A. D. (2001). Iron deficiency and chlorosis in orchard and vineyard ecosystems. European Journal of Agronomy, 15(2), 71-92.

Wahba, M., Fawkia, L. A. B. İ. B., & Zaghloul, A. (2019). Management of calcareous soils in arid region. International Journal of Environmental Pollution and Environmental Modelling, 2(5), 248-258.

Wahid, F., Fahad, S., Danish, S., Adnan, M., Yue, Z., Saud, S., ... & Datta, R. (2020). Sustainable management with mycorrhizae and phosphate solubilizing bacteria for enhanced phosphorus uptake in calcareous soils. Agriculture, 10(8), 334.

Zhang, Y., Zhang, S., Wang, R., Cai, J., Zhang, Y., Li, H., ... & Jiang, Y. (2016). Impacts of fertilization practices on pH and the pH buffering capacity of calcareous soil. Soil Science and Plant Nutrition, 62(5-6), 432-439.

7

Waterlogged Soil Challenges and Strategies

Introduction

Waterlogging refers to the accumulation of excessive water in the root zone, leading to anaerobic conditions. This surplus water hampers the exchange of gases with the atmosphere, and biological processes deplete the oxygen supply in both soil and water, resulting in anaerobiosis, anoxia, or oxygen deficiency. In India, approximately 11.6 million hectares, constituting 8.3% of its net sown area, are affected by waterlogging (Planning Commission, 2011). According to Brundtland and Khalid (1987), available estimates indicate an annual global loss of 1.5 million hectares of irrigated land due to salinity and waterlogging. Regrettably, data regarding the occurrence and extent of these issues are inconsistent and incomplete. More recent estimates, provided by Datta and Joshi (1992), range from 5.5 million to 13 million hectares. After conducting a comprehensive global survey, it has been determined that submerged and waterlogged soils cover approximately 5 to 7% of the Earth's land surface. The total area of waterlogged soil worldwide is estimated to be around 700 to 1000 million hectares. Tropical swamps, rice fields, and floodplains collectively represent nearly 14%, 12%, and 10% of the total waterlogged area, respectively. In India, Odisha, West Bengal, Bihar, and Uttar Pradesh have the highest concentration of waterlogged soil, with an estimated total area of one million hectares. The eastern region bears a significant portion of this burden, with over 20% of the affected land suffering from surface waterlogging, which severely diminishes productivity. Waterlogging is characterized by the soil becoming unproductive and infertile due to excess moisture, creating anaerobic conditions known as waterlogged soils. Waterlogged soil presents a significant challenge to agricultural productivity and soil health, characterized by excessive water saturation that restricts oxygen availability to plant roots. In this chapter, we delve into the multifaceted aspects of waterlogged soil, exploring its causes, defining characteristics, and far-reaching consequences. The serves as a comprehensive overview, underscoring the detrimental impact of waterlogging on agriculture, ecosystems, and human well-being. It identifies key factors contributing to waterlogged soil, including heavy rainfall, inadequate drainage, high water tables, over-irrigation practices, and natural landscape depressions. These elements collectively exacerbate soil saturation, transforming fertile land into inhospitable terrain for plant growth. the text elucidates the defining features of waterlogged soil, emphasizing the diminished oxygen levels, compromised soil structure, heightened soil acidity, nutrient imbalances, and increased vulnerability to erosion. These characteristics not only impede plant growth but also disrupt the delicate balance of soil ecosystems, leading to cascading ecological repercussions. Moreover, the underscores the dire consequences of waterlogged soil, ranging from diminished crop yields and loss of soil structure to heightened disease incidence and decreased biodiversity. Such outcomes reverberate across agricultural landscapes, affecting food security, environmental resilience, and ecosystem services. These strategies encompass improving drainage infrastructure, implementing contour farming techniques, incorporating organic matter into the soil, selecting resilient crop varieties, and adopting prudent irrigation practices. By addressing the root causes and symptoms of waterlogging, these management approaches seek to restore soil health, enhance agricultural sustainability, and mitigate the adverse impacts of soil saturation on human livelihoods and ecosystem functioning.

Figure 1: Waterlogging in crop

Causes of Waterlogged Soil: Waterlogged soil can result from various factors, including:

1. **Excessive Rainfall:** A Deluge of Consequences: Heavy or prolonged rainfall, a blessing in moderation, can metamorphose into a curse when it inundates the soil beyond its capacity. In poorly drained areas, where the soil's natural drainage prowess is compromised, each raindrop becomes a harbinger of saturation. The deluge permeates the soil matrix,

saturating every pore and crevice, until the ground itself relinquishes its absorbent capacity. What was once a nurturing rain becoming a torrential downpour, overwhelming the land and drowning the hopes of farmers and ecosystems alike. The consequences of excessive rainfall reverberate far and wide, from agricultural fields rendered impassable by waterlogged soils to urban landscapes besieged by floods. With each rainfall event, the soil's capacity to absorb water is tested, and in the crucible of saturation, its vulnerabilities are laid bare. The excess water, unable to find respite in the depths of the earth or the embrace of drainage systems, accumulates on the surface, transforming fertile fields into waterlogged wastelands.

2. **Poor Drainage:** Unraveling the Constrictions: Compacted soil, impermeable layers, & inadequate drainage systems conspire to shackle the natural flow of water, confining it to the surface or relegating it to stagnant pools. The intricate network of soil pores, a conduit for the passage of water and air, finds itself ensnared by the encroaching tendrils of compaction. Impermeable layers, whether natural or man-made, serve as barriers to infiltration, diverting the flow of water and exacerbating soil saturation. Inadequate drainage systems, ill-suited to the task at hand, falter under the weight of excess water, unable to provide the escape route that saturated soils so desperately seek. Ditches become stagnant reservoirs, tile drains clogged arteries, and French drains mere conduits of frustration. With each rainfall event, the shortcomings of poor drainage are laid bare, as waterlogged soils bear witness to the folly of neglecting nature's hydrological imperatives.

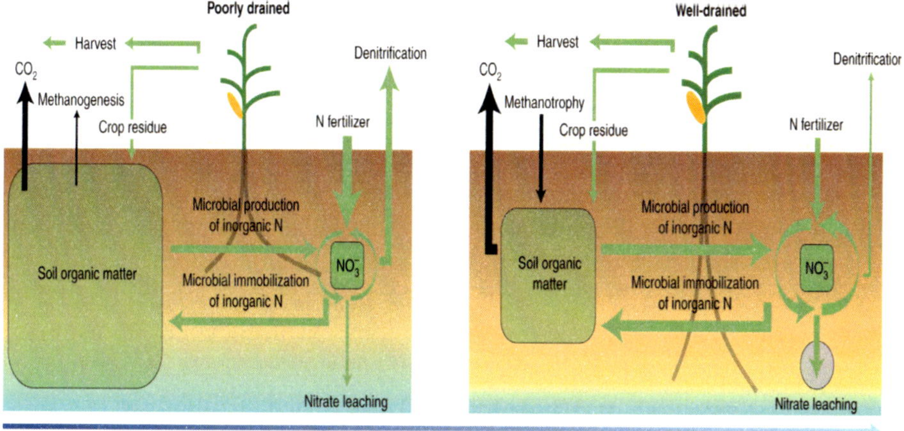

Figure 2: Drainage system

3. **High Water Table:** Plumbing the Depths of Saturation: In regions blessed—or cursed—with shallow groundwater, the delicate equilibrium of the water table hangs in the balance, swayed by the capricious whims of nature and human intervention alike. Excessive irrigation or rainfall, cascading through the soil profile, percolates downward, replenishing the subterranean reservoirs that lie beneath. With each influx of water, the water table rises, inching ever closer to the surface, until the line between soil and water blurs into obscurity. The consequences of a high-water table extend far beyond the confines of the soil, permeating the very fabric of ecosystems and human endeavors. Wetlands swell with newfound vitality, as the life-giving waters of the aquifer rise to meet them. Yet, in the realm of agriculture, the rising tide of the water table spells disaster, inundating fields and drowning the hopes of farmers in a deluge of despair.
4. **Over-irrigation:** The Drowning Depths of Excess: In the quest for abundance, the line between sufficiency and excess becomes blurred, as water, the elixir of life, transforms from boon to bane. Over-irrigation, a misguided attempt to quench the thirst of crops, inundates the soil with a torrent of water, overwhelming its capacity to absorb and drain. Each drop, a drop too many, becomes a harbinger of saturation, as the soil surrenders to the relentless onslaught. The consequences of over-irrigation reverberate far beyond the boundaries of the field, infiltrating the very heart of ecosystems and communities alike. Wetlands shrink, deprived of the life-giving waters they depend upon, while rivers swell with the excesses of human intervention. In the crucible of saturation, the delicate balance of soil ecology is upended, as anaerobic conditions prevail and the roots of plants suffocate in a watery grave.
5. **Natural Depressions:** The Abyss of Accumulation: In the undulating landscape of the earth, depressions carve out niches of stagnation, where water finds respite from the relentless pull of gravity. Low-lying areas, nestled in the embrace of hills and valleys, become repositories of precipitation, accumulating water until the soil itself relinquishes its absorbent capacity. What was once a gentle slope becoming a watery abyss, as the forces of erosion give rise to pools and puddles. The consequences of natural depressions extend beyond the confines of the landscape, permeating the very fabric of ecosystems and human settlements alike. Wetlands thrive in the shelter of these depressions, teeming with life and vitality, while agricultural fields languish in the

suffocating embrace of waterlogged soils. In the intricate dance of water and land, natural depressions stand as monuments to the power and peril of precipitation.

Characteristics of Waterlogged Soil: Waterlogged soil exhibits several characteristic features, including:

1. **Reduced Oxygen Availability:** A Suffocating Dilemma: Within the waterlogged soil, a silent battle unfolds as oxygen, the lifeblood of root respiration, struggles to penetrate the suffocating embrace of saturation. Hindered by the impenetrable barrier of excess water, oxygen diffusion falters, plunging the soil into a realm of hypoxic or anoxic conditions. In this oxygen-deprived environment, plant roots gasp for breath, their vital functions stifled by the absence of this essential element. The consequences of reduced oxygen availability reverberate throughout the soil ecosystem, as anaerobic microorganisms thrive in the absence of their aerobic counterparts. With each passing moment, the delicate balance of soil ecology is upended, as the forces of decomposition and nutrient cycling are reshaped by the suffocating veil of saturation. In the crucible of hypoxia, plant roots wither and decay, their once-vibrant vitality extinguished by the relentless march of anaerobicity.
2. **Poor Soil Structure:** A Compacted Conundrum: Excess water, once a source of nourishment and vitality, becomes a harbinger of compaction as soil particles succumb to the weight of saturation. Pore spaces, once the conduits of water and air, become constricted as particles are compressed, reducing the soil to a dense and impermeable matrix. In this compacted conundrum, roots struggle to navigate the labyrinth of soil structure, their growth stunted by the lack of space and aeration. The consequences of poor soil structure extend far beyond the confines of the soil profile, permeating the very fabric of agricultural productivity and ecosystem resilience. With each passing season, the soil loses its capacity to absorb and retain water, leaving crops vulnerable to the vagaries of drought and inundation alike. In the crucible of compaction, the delicate balance of soil fertility is eroded, as the forces of erosion and degradation chip away at the foundation of agricultural sustainability.
3. **Increased Soil Acidity:** The Souring of Soil pH Amidst the anaerobic depths of waterlogged soil, a silent transformation unfolds as acids, unleashed by microbial metabolism, permeate the soil matrix. Anaerobic conditions, a breeding ground for acid-producing microorganisms, promote the release of acids into the soil solution, lowering the pH and souring the very essence of soil fertility. In this acidic milieu,

plant roots struggle to absorb essential nutrients, their uptake hindered by the hostile environment that surrounds them. The consequences of increased soil acidity reverberate throughout the soil ecosystem, as the delicate balance of nutrient availability is upended by the souring of soil pH with each passing season, the soil becomes less hospitable to plant life, as essential nutrients are locked away in insoluble forms, beyond the reach of root uptake. In the crucible of acidity, the very foundation of agricultural productivity is undermined, as crops languish in the suffocating embrace of waterlogged soils.

4. **Nutrient Imbalances:** The waterlogged soil, a subtle shift occurs as the delicate balance of soil ecology is reshaped by the suffocating veil of saturation. Altered microbial activity, spurred by anaerobic conditions, disrupts the delicate equilibrium of nutrient cycling, leading to a cascade of consequences for plant health and productivity. With each passing moment, essential nutrients are mobilized or immobilized, their availability shaped by the capricious whims of microbial metabolism. The consequences of nutrient imbalances reverberate throughout the soil ecosystem, as the delicate dance of nutrient cycling is disrupted by the suffocating embrace of saturation. With each passing season, crops struggle to meet their nutrient requirements, their growth stunted by deficiencies or toxicity arising from the altered soil environment. In the crucible of imbalance, the very foundation of agricultural sustainability is threatened, as crops languish in the nutrient-deprived wasteland of waterlogged soils.
5. **Increased Susceptibility to Erosion:** The Eroding Edges of Saturation: Saturated soil, once a bastion of stability and fertility, becomes a vulnerable frontier as the forces of erosion seize the opportunity presented by saturation. With each rainfall event, water becomes a relentless agent of erosion, scouring the soil surface and carrying away precious topsoil in its inexorable flow. In the crucible of saturation, the delicate balance of soil structure is eroded, leaving behind a barren wasteland of depleted fertility. The consequences of increased susceptibility to erosion extend far beyond the confines of the soil profile, permeating the very fabric of agricultural productivity and environmental resilience. With each passing season, the soil loses its capacity to support plant life, as essential nutrients are stripped away and fertile topsoil is carried downstream. In the crucible of erosion, the very foundation of agricultural sustainability is threatened, as the forces of degradation chip away at the fertile substrate upon which our food security depends.

Consequences of Waterlogged Soil: The consequences of waterlogged soil extend beyond plant health and productivity, impacting various aspects of soil and ecosystem functioning:

1. **Reduced Crop Yields:** Waterlogged soil, once a nurturing medium for plant growth, transforms into a barrier to productivity as root growth and nutrient uptake are stifled by its suffocating embrace. In this waterlogged realm, roots struggle to penetrate the compacted soil, their growth stunted by the lack of oxygen and the impenetrable barrier of saturation. As nutrient uptake falters, plants languish in a state of stunted growth, their potential for yield hampered by the adverse conditions that surround them. The consequences of reduced crop yields reverberate throughout the agricultural landscape, as farmers grapple with diminished returns and uncertain futures. With each passing season, the soil becomes less hospitable to plant life, as waterlogged conditions rob crops of the nutrients and oxygen they need to thrive. In the crucible of saturation, the promise of abundance gives way to the reality of scarcity, as crop yields fall short of expectations and food security hangs in the balance.
2. **Loss of Soil Structure:** Compaction and anaerobic conditions conspire to erode the very foundation of soil structure, leaving behind a barren wasteland devoid of fertility and resilience. In waterlogged soil, the delicate balance of soil particles is disrupted by the weight of saturation, as compaction compresses the soil matrix and reduces its capacity to support plant growth and retain water. With each passing season, the soil loses its ability to absorb and store water, leaving crops vulnerable to the vagaries of drought and inundation alike. The consequences of loss of soil structure extend far beyond the boundaries of the field, permeating the very fabric of agricultural productivity and environmental resilience. With each passing season, the soil becomes less hospitable to plant life, as essential nutrients are leached away and fertile topsoil is carried downstream. In the crucible of erosion, the very foundation of agricultural sustainability is threatened, as the forces of degradation chip away at the fertile substrate upon which our food security depends.
3. **Increased Disease Incidence**: Waterlogged soil, with its anaerobic depths and stagnant pools, becomes a breeding ground for soil-borne pathogens, increasing the risk of disease outbreaks in agricultural crops. In this waterlogged realm, pathogens find refuge from the dry conditions that inhibit their growth, proliferating in the moist and oxygen-deprived environment that surrounds them. With each passing season, the risk of disease incidence rises, as crops succumb to the relentless assault of

fungal and bacterial pathogens. The consequences of increased disease incidence reverberate throughout the agricultural landscape, as farmers grapple with diminished yields and increased production costs. With each passing season, the soil becomes less hospitable to plant life, as pathogens proliferate unchecked and crop losses mount. In the crucible of saturation, the promise of abundance gives way to the reality of scarcity, as disease outbreaks threaten to undermine the very foundation of agricultural productivity.

4. **Decreased Soil Biodiversity:** Anaerobic conditions in waterlogged soil serve as a barrier to soil biodiversity, suppressing the activity of beneficial soil organisms and leading to a decline in overall diversity. In this oxygen-deprived environment, aerobic organisms struggle to survive, while anaerobic organisms thrive in the absence of competition. With each passing season, the delicate balance of soil ecology is upended, as the forces of decomposition and nutrient cycling are reshaped by the suffocating veil of saturation. The consequences of decreased soil biodiversity extend far beyond the confines of the field, permeating the very fabric of agricultural productivity and environmental resilience. With each passing season, the soil becomes less hospitable to plant life, as essential nutrients are locked away in insoluble forms, beyond the reach of root uptake. In the crucible of imbalance, the very foundation of agricultural sustainability is threatened, as crops languish in the nutrient-deprived wasteland of waterlogged soils.
5. **Loss of Ecosystem Services:** Waterlogged soil, once a bastion of fertility and resilience, becomes a barrier to the provision of ecosystem services as soil fertility, water filtration, and carbon sequestration are impaired by its suffocating embrace. In this waterlogged realm, the delicate balance of soil ecology is disrupted by the weight of saturation, as essential services are compromised by the adverse conditions that surround them. With each passing season, the promise of abundance gives way to the reality of scarcity, as the forces of degradation chip away at the very foundation of ecosystem vitality. The consequences of loss of ecosystem services extend far beyond the confines of the field, permeating the very fabric of agricultural productivity and environmental resilience. With each passing season, the soil becomes less hospitable to plant life, as waterlogged conditions rob crops of the nutrients and oxygen they need to thrive. In the crucible of saturation, the promise of abundance gives way to the reality of scarcity, as ecosystem services are compromised and the delicate balance of soil ecology is upended.

Managing waterlogged soil

1. **Improving Drainage:** Addressing poor drainage is paramount to alleviating waterlogged soil conditions. Installation of drainage systems, including ditches, tile drains, or French drains, provides avenues for excess water removal, allowing the soil to breathe once more. By facilitating the swift evacuation of surplus water, these interventions mitigate the risk of saturation, restoring equilibrium to the soil-water balance. Installing a subsurface drainage system facilitates the management of water table levels and the desalination of soils through leaching, utilizing either irrigation water or monsoon rains. Operational research conducted on the Sampla project demonstrated a rapid reduction in surface layer salinity following drainage, decreasing from approximately 50 dS m^{-1} in 1984 to about 5 dS m^{-1} in 1991 (Rao *et al.*, 1991). In numerous small-scale pilot projects across Haryana State, predominantly managed by farmers, the immediate impacts of subsurface drainage were significant. These included a notable rise in cropping intensity, a shift towards more profitable crops in the cropping pattern, a remarkable boost in crop yields, and enhanced fertilizer productivity. These collective changes resulted in a substantial upsurge in farm incomes. Nonetheless, it's imperative to approach the outcomes of these pilot areas with caution. The sample size of farms was relatively small due to the modest size of the study areas, and with the exception of Ismila, data collection occurred over a relatively brief period.
2. **Implementing Contour Farming:** Contour farming techniques offer a strategic approach to waterlogged soil management, particularly in sloping areas prone to runoff and erosion. Terracing or contour plowing aligns cultivation practices with the natural contours of the land, reducing the velocity of water runoff and enhancing soil water retention. By curbing erosion and minimizing surface runoff, contour farming mitigates the risk of waterlogging, fostering soil stability and resilience.
3. **Applying Organic Matter:** Incorporating organic matter, such as compost or mulch, into waterlogged soil serves as a transformative intervention to enhance soil structure and infiltration rates. Organic amendments replenish soil organic carbon, fostering microbial activity and improving soil aggregation. As soil structure is revitalized, water infiltration rates increase, reducing the likelihood of waterlogging and promoting aeration within the soil profile.

4. **Selecting Appropriate Crops:** Strategic crop selection plays a pivotal role in managing waterlogged soil, with an emphasis on varieties adapted to such conditions or alternative cropping systems resilient to saturation. Opting for waterlogging-tolerant crop varieties or exploring innovative systems like rice-fish or wetland agriculture allows for sustainable production in waterlogged environments. By aligning crop choices with soil hydrology, farmers can mitigate the impact of waterlogging and harness the potential of waterlogged lands.
5. **Avoiding Over-irrigation:** Effective irrigation management is essential in preventing waterlogging and optimizing water use efficiency. By adopting proper scheduling and irrigation practices tailored to crop water requirements and soil characteristics, excessive water application can be avoided. Matching irrigation rates to soil infiltration rates ensures that water uptake by plants aligns with soil water-holding capacity, mitigating the risk of saturation and promoting sustainable water management practices. In concert, these strategies offer a comprehensive framework for managing waterlogged soil, safeguarding agricultural productivity, and fostering soil health in the face of hydrological challenges. Through proactive intervention and strategic planning, the resilience of waterlogged landscapes can be enhanced, ensuring the sustainable utilization of land resources for generations to come.

Conclusion

Waterlogged soil poses significant challenges to agriculture, ecosystems, and infrastructure, affecting plant growth, soil health, and land usability. However, through effective strategies such as improved drainage systems, sustainable water management, crop adaptation, and soil amendments, these challenges can be mitigated. Implementing these solutions not only enhances productivity but also promotes environmental sustainability. Addressing waterlogging requires a combination of traditional knowledge and modern innovation to ensure resilient land use practices for the future. Waterlogged soil poses multifaceted challenges to agricultural sustainability and ecosystem resilience. Effective management strategies are imperative to mitigate its adverse impacts and restore soil health. Improving drainage infrastructure, implementing contour farming techniques, applying organic matter, selecting appropriate crop varieties, and avoiding over-irrigation are critical interventions to address waterlogging. While pilot projects have shown promising results in enhancing agricultural productivity and farm incomes, further research and implementation are warranted to scale up these interventions and ensure their long-term effectiveness. By adopting a holistic approach to waterlogged soil management, we can safeguard agricultural livelihoods, promote environmental sustainability, and secure food security for present and future generations.

References

Brundtland, G.M., Khalid, M., 1987. Our Common Future: Report of the World Commission on the Environment. UNEP Governing Council, Oxford University Press, Oxford.

Buresh, R. J., Ramesh Reddy, K., & Van Kessel, C. (2008). Nitrogen transformations in submerged soils. Nitrogen in agricultural systems, 49, 401-436.

Erich, E., Drohan, P. J., Ellis, L. R., Collins, M. E., Payne, M., & Surabian, D. (2010). Subaqueous soils: their genesis and importance in ecosystem management. Soil use and management, 26(3), 245-252.

Kamra, S. K. (2015). An overview of subsurface drainage for management of waterlogged saline soils of India. Water and Energy International, 58(6), 46-53.

Kirk, G. (2004). The biogeochemistry of submerged soils. John Wiley & Sons.

Lakshmi, C. S. R., Rao, P. C., Sreelatha, T., Padmaja, G., Madhavi, M., Rao, P. V., & Sireesha, A. (2014). Biochemical changes in submerged rice soil amended with different vermicomposts under integrated nutrient management. Journal of the Indian Society of Soil Science, 62(2), 131-139.

Manghwar, H., Hussain, A., Alam, I., Khoso, M. A., Ali, Q., & Liu, F. (2024). Waterlogging stress in plants: Unraveling the mechanisms and impacts on growth, development, and productivity. Environmental and Experimental Botany, 105824.

Marschner, P. (2021). Processes in submerged soils–linking redox potential, soil organic matter turnover and plants to nutrient cycling. Plant and Soil, 464(1), 1-12.

Osman, K. T. (2018). Management of soil problems (pp. 255-298). Cham, Switzerland: Springer International Publishing.

Pan, J., Sharif, R., Xu, X., & Chen, X. (2021). Mechanisms of waterlogging tolerance in plants: Research progress and prospects. Frontiers in Plant Science, 11, 627331.

Panda, D. (2005). Chemistry of nitrogen transformation in submerged soil and scientific management of urea fertilizer for higher rice productivity. Journal of the Indian Society of Soil Science, 53(4), 500-513.

Qamar, R., Shafaat, S., & Javeed, H. M. R. (2023). Management of Crops in Water-Logged Soil. In Disaster Risk Reduction in Agriculture (pp. 233-275). Singapore: Springer Nature Singapore.

Rao, S. N., Prasad, Y. V. S. N., & Shetty, M. D. (1991). The behaviour of model screw piles in cohesive soils. Soils and Foundations, 31(2), 35-50.

Sahrawat, K. L. (2004). Organic matter accumulation in submerged soils. Advances in agronomy, 81, 169-201.

Singh, S. H. I. K. H. A., Singh, S., Singh, G., & Kaur, G. (2022). Soil waterlogging: cause, impact, and management. Soil Constraints on Crop Production, 62.

Valipour, M. (2014). Drainage, waterlogging, and salinity. Archives of Agronomy and Soil Science, 60(12), 1625-1640.

Zhao, J., Ma, H., Yan, H., Jiang, T., & Zhu, W. (2024). Management of waterlogged area based on a three-dimensional agricultural model of ponds and dry land. Physics of Fluids, 36(7).

8

Soil Crusting and Their Management

Introduction

Soil crusts are surface alterations in the topsoil resulting from natural processes, such as the impact of raindrops followed by drying. These crusts manifest as thin, hardened layers on the soil surface, particularly common in arid and semi-arid regions. Their thickness typically varies from less than 1 mm to 5 cm (Evans and Boul, 1968). Once dry, these layers become denser, harder, and more brittle than the underlying soil, reducing both the number and size of pores and reorganizing the soil's pore structure. In temperate regions, surface crusts primarily form on unstable loamy soils, particularly in areas under cultivation. In tropical regions, soil crusting affects a broader range of soils and poses a significant problem not only in arid zones but across various climatic conditions. In humid areas, intensive farming exposes the soil surface to high-energy rainfall for extended periods, accelerating the degradation of soil organic matter. As a result, bare soil areas expand, and biomass production declines. From an agronomic perspective, the most critical drawbacks of soil crusting are its negative effects on seedling emergence and water infiltration.

Seedling emergence is particularly affected for crops with small seeds or when the timing of emergence is crucial due to climate or market demands. In some instances, costly replanting becomes necessary. The reduction in water infiltration caused by crusted soils creates significant challenges for irrigation, especially in areas where water scarcity limits large-scale use, forcing farmers to adopt more efficient water management strategies. Reduced water infiltration also leads to increased surface runoff, contributing to soil erosion and environmental risks, such as the pollution of surface waters due to nutrient loss. Additionally, restricted gas exchange between the soil and atmosphere further contributes to lower crop yields.

Epstein and Grant (1973) observed that soil erodibility is closely linked to the rate and extent of crust formation, with soil loss peaking within the first 10 minutes of rainfall before stabilizing. It is well established that during crust-forming rain events, water infiltration declines over time, leading to an increase in runoff volume. This increase in runoff, combined with the intensified energy of raindrops due to overland flow, raises the overall erosivity of the rain event. However, the simultaneous decrease in soil erosion, despite higher water erosivity, suggests that crust formation results in reduced soil erodibility over time.

Formation of Soil Crust

Soil crusting is a widespread issue affecting agricultural productivity, water infiltration, and seedling emergence. Understanding the detailed formation process of soil crusts is crucial for devising strategies to mitigate their negative impacts. Soil crusts primarily form due to the breakdown of soil aggregates, redistribution of fine particles, and subsequent surface compaction, influenced by physical, chemical, and biological factors. Below is a step-by-step detailed explanation of the formation of soil crusts for inclusion in a book chapter.

1. Initiation Aggregate Breakdown by Raindrop Impact

The formation of soil crusts begins with the breakdown of soil aggregates, a process mainly driven by the kinetic energy exerted by raindrop impact or irrigation water on exposed soil surfaces. When raindrops fall on unprotected soil, they exert substantial pressure on the soil particles, disrupting the aggregates that are held together by weak bonds. Aggregates are composed of a mix of clay, silt, sand, and organic materials, and the breakdown occurs more easily in soils with lower organic matter content or where soil structure is inherently weak. The energy from raindrop impact can cause a variety of disturbances, including:

Disaggregation: Soil aggregates disintegrate, releasing finer particles (silt and clay) and organic matter.

Soil particle redistribution: The water from raindrops causes these finer particles to be suspended and transported over short distances across the soil surface. The severity of this initial breakdown is influenced by several factors:

Soil texture: Soils with higher clay and silt content are more prone to aggregate breakdown than sandy soils, which are better drained.

Organic matter content: Organic matter acts as a binding agent for soil particles. Soils have low in organic matter tend to lose structure more readily, leading to faster breakdown.

Vegetation cover: Soils that lack protective vegetation or mulch are more vulnerable to raindrop impact, as plant canopies and ground cover buffer the energy of raindrops. This process of soil particle disintegration is also referred to as slaking, which describes the rapid breakdown of aggregates upon sudden wetting.

2. Dispersion of Fine Particles

Once soil aggregates are broken down by the impact of rain or irrigation, the finer particles, particularly silt and clay, become suspended in the thin layer of water on the soil surface. The behavior of these particles during wetting is critical to the crust formation process.

Clay minerals: The type of clay minerals present in the soil plays a major role in dispersion. For instance, montmorillonite (a type of smectite) tends to swell and disperse easily, while kaolinite is more stable. Soils rich in smectitic clays are more susceptible to dispersion, leading to the formation of denser crusts.

Soil sodicity: High levels of sodium ions in the soil enhance the dispersibility of clay particles by reducing the attractive forces between them. Sodic soils are particularly prone to crust formation due to the breakdown of soil structure upon wetting.

Dispersion occurs when clay particles, which carry a negative charge, repel each other under wet conditions. As a result, these particles remain suspended in water for extended periods before settling. This creates an even distribution of fine particles on the soil surface, filling the voids and pores.

3. Formation of a Thin Dense Layer

As the dispersed fine particles settle, they migrate downward and begin to clog the soil pore at the surface. This process, referred to as surface sealing, creates a thin, compacted layer on the soil surface that is rich in fine particles like clay and silt. The sealing layer significantly alters the soil's physical structure by reducing the size and connectivity of surface pores, thereby limiting water infiltration and gas exchange

Key mechanisms involved in surface sealing include:

Pore clogging: The settling fine particles block the macrospores (larger pores) of the soil, leading to a dramatic reduction in permeability.

Compaction: The surface layer becomes denser as water evaporates, leaving behind a more compact and less porous structure. This compaction is especially prevalent in soils with higher clay content, where the fine particles form tight bonds.

At this stage, the soil surface has transformed from a relatively loose structure to a much denser, less permeable crust. The formation of the sealed layer can occur rapidly, often within a few minutes of rain or irrigation, particularly in soils with fine-textured particles.

4. Drying and Hardening of the Crust

After the formation of the surface seal, the soil begins to dry as water is either absorbed into the soil below or evaporates from the surface. As drying occurs, the fine particles in the sealing layer consolidate further, leading to a hardened crust. This hardening is particularly pronounced in soils rich in expansive clays, which shrink as they dry, causing the surface layer to become compact and impervious to water.

Several factors influence the drying and hardening process:

Clay mineralogy: Soils with expansive clay minerals like smectite experience significant shrinkage and consolidation upon drying. These soils form thicker, harder crusts that are difficult for water and roots to penetrate. In contrast, soils with kaolinitic clays may form thinner crusts that crack and break apart upon drying.

Water content: The amount of water present in the soil during drying also affects the severity of crust formation. Rapid drying can lead to more pronounced compaction and hardening of the crust.

Evaporation rate: In arid and semi-arid environments, high evaporation rates accelerate crust formation, leaving behind a hard, dense layer more quickly.

As a result of this drying and hardening, the surface becomes resistant to infiltration during future rainfall events, exacerbating runoff and erosion issues. This hardened layer restricts the movement of water, air, and roots into the soil, which significantly affects seedling emergence and plant growth.

5. Reinforcement by Repeated Wetting and Drying Cycles

Soil crusts can become progressively stronger and more compact through repeated cycles of wetting and drying. Each time the soil is wetted, additional breakdown of aggregates and further dispersion of fine particles occur, reinforcing the existing crust. When the soil dries again, the newly settled particles add to the thickness and hardness of the crust layer.

Repeated raindrop impact: Continued rainfall or irrigation on already crusted soils can worsen the problem by breaking apart any remaining aggregates and creating a thicker, more resilient crust.

Soil erosion: In some cases, runoff from rainfall on crusted soils can erode the unprotected surface, removing valuable topsoil and leaving behind a more compacted, erosion-resistant layer.

6. Influence of Biological Factors

Biological activity plays a significant role in either mitigating or promoting soil crust formation. In healthy soils, biological organisms such as plants, fungi, and microorganisms contribute to soil aggregation and the stabilization of the soil surface.

Soil Organic Matter (SOM): Organic matter is a key binding agent in the formation of stable soil aggregates. Soils with higher organic matter content are less prone to crust formation because the organic compounds (humic substances, roots, microbial secretions) help maintain aggregate structure even during wetting. The presence of organic matter reduces the dispersibility of fine particles and improves water infiltration.

Microbial activity: Microorganisms, such as fungi and bacteria, produce polysaccharides that act as "glue" to bind soil particles together. These biological binding agents enhance aggregate stability and reduce the risk of physical crust formation. Additionally, cyanobacteria and algae in arid and semi-arid regions can form biological crusts that stabilize the soil surface.

Plant roots and vegetation cover: The presence of vegetation helps protect the soil from raindrop impact, reducing the likelihood of aggregate breakdown. Plant roots also create channels that improve water infiltration and contribute to the structural integrity of the soil. Soils without vegetation cover are much more susceptible to crust formation due to the lack of protective biomass.

7. Environmental Factors Contributing to Crust Formation

Environmental conditions also play a crucial role in soil crust formation. The following factors significantly influence the severity and persistence of soil crusts:

Rainfall intensity: Heavy and intense rainfall events are more likely to cause rapid aggregate breakdown and fine particle dispersion, leading to faster crust formation. Gentle rains, by contrast, are less disruptive and allow for better infiltration.

Topography: Sloped or uneven terrain can exacerbate surface runoff, leading to erosion and further crust development. Soils on flat terrain may have a higher risk of crusting due to standing water that promotes fine particle settling.

Climate: Arid and semi-arid regions, where rainfall is infrequent but intense, are particularly prone to soil crust formation. High evaporation rates in these regions also contribute to rapid drying and hardening of the crust.

Classification of Soil Crusts

Soil crusts can be classified based on their formation processes, physical characteristics, and biological activity. The following are the main types of soil crusts:

1. Physical Crusts

Physical or structural crusts form primarily due to the breakdown of soil aggregates and the rearrangement of soil particles caused by external forces such as raindrop impact or irrigation. These crusts are subdivided into:

a) **Structural Crusts:** Structural crusts develop when raindrops or irrigation disrupt soil aggregates, leading to the migration and deposition of fine particles (silt and clay) on the soil surface. As the surface dries, these fine particles compact and harden, forming a dense, low-porosity layer. These crusts are most common in soils with low organic matter and high silt or clay content.

b) **Depositional Crusts:** Depositional crusts form when water flow (either from rainfall or irrigation) moves fine particles across the soil surface, which then settle and consolidate into a crust as the water evaporates. These crusts are typically found on sloped or uneven terrains where surface runoff occurs. The transported materials fill soil pores, reducing infiltration and further compacting the surface.

c) **Erosional Crusts:** Erosional crusts result from the physical removal of soil particles due to wind or water erosion. The remaining soil forms a compacted, hardened layer as it dries. These crusts are common in regions with high erosion rates, such as semi-arid or windy areas, where topsoil is eroded, leaving behind a resistant layer.

2. Biological Crusts

Biological soil crusts, also called bio crusts, are formed by the activity of living organisms on the soil surface. These crusts typically consist of communities of cyanobacteria, algae, fungi, lichens, mosses, and other microorganisms that bind soil particles together and create a stable surface layer. Biological crusts are most common in arid and semi-arid regions, but they can also occur in various ecosystems.

a) **Microbial Crusts:** Microbial crusts are dominated by microorganisms such as cyanobacteria and algae, which secrete extracellular polymeric substances (EPS) that bind soil particles together. These organisms thrive in moisture and form a protective layer that stabilizes the soil, improving its resistance to erosion. These crusts often contribute to nitrogen fixation and nutrient cycling in the soil.

b) **Lichen and Moss Crusts:** Lichens and mosses can also form crusts by covering the soil surface with their structures, which protect the underlying soil from erosion and raindrop impact. These crusts are thicker and more cohesive than microbial crusts and are commonly found in deserts or regions with poor vegetation. Lichen and moss crusts improve soil stability but may limit seedling emergence by creating a dense barrier.

3. Chemical Crusts

Chemical crusts form due to the precipitation of salts or minerals on the soil surface. These crusts are particularly common in saline or sodic soils and develop when dissolved salts move upward through the soil profile via capillary action. As the water evaporates, salts are left behind, creating a hard, impermeable surface layer.

a) **Saline Crusts:** Saline crusts occur in soils with high salt concentrations, particularly in regions with poor drainage or high evaporation rates. The salts precipitate and accumulate at the surface, leading to a hard, dense layer that impedes water infiltration and seedling growth. Saline crusts are most common in arid and semi-arid regions.

b) **Gypsum Crusts:** In soils rich in calcium sulfate (gypsum), gypsum crusts can form when the mineral precipitates and accumulates at the soil surface. These crusts are more common in soils with low organic matter and limited biological activity. They can be less dense than saline crusts but still reduce infiltration and contribute to soil surface compaction.

4. Rain-Splash Crusts

Rain-splash crusts form when the energy from raindrops displaces soil particles, causing the finer particles to settle and consolidate into a thin, hard layer. This type of crust typically forms immediately after rainstorms, especially in soils with a loose, unprotected surface. These crusts can be transient, often breaking apart as the soil dries or experiences disturbance.

5. Composite Crusts

Composite crusts result from a combination of the above processes, involving both physical and biological factors. For instance, a physical crust may form initially from raindrop impact, but over time, biological activity may stabilize and modify the crust, integrating organic materials and microbial life. These mixed crusts can be more resistant to disturbance and can vary significantly in thickness and structure.

How to improve of Soil Crusting

Soil crusting is a significant challenge for soil health and agricultural productivity, primarily affecting water infiltration, seedling emergence, and root growth. Managing soil crusting requires an integrated approach that addresses the physical, chemical, and biological factors that contribute to its formation. Several management practices can help prevent crust formation or mitigate its effects. These practices aim to maintain or improve soil structure, enhance organic matter content, and promote biological activity. Below is a detailed overview of strategies for managing soil crusting.

1. Increasing Organic Matter Content

One of the most effective ways to prevent soil crusting is by increasing the organic matter content in the soil. Organic matter improves soil structure, promotes aggregate stability, and enhances water infiltration.

Adding organic amendments: Applying organic materials such as compost, manure, crop residues, or green manure to the soil increases the level of soil organic matter (SOM). Organic matter acts as a binding agent for soil particles, helping to stabilize aggregates and reduce the likelihood of crust formation.

Incorporating cover crops: Growing cover crops, such as legumes or grasses, during fallow periods can increase organic matter in the soil. Cover crops also protect the soil surface from raindrop impact and help maintain soil structure through their root systems.

Mulching: Applying mulch to the soil surface can protect it from the impact of raindrops, reduce evaporation, and keep the soil moist. Organic mulches such as straw, leaves, or crop residues decompose over time, adding organic matter to the soil and improving aggregate stability.

2. Minimizing Soil Disturbance through Conservation Tillage

Conservation tillage practices aim to reduce soil disturbance, thereby maintaining soil structure and preventing the breakdown of aggregates that can lead to crust formation.

No-till or reduced-till systems: In no-till or reduced-till systems, the soil is left undisturbed, allowing natural processes to improve soil structure over time. These systems help preserve soil organic matter, reduce the risk of aggregate breakdown, and protect the soil surface from erosion and compaction.

Strip-till: In strip-till systems, tillage is limited to narrow strips where seeds will be planted, leaving the rest of the soil undisturbed. This approach reduces soil disturbance while maintaining the protective cover of plant residues on the soil surface, which reduces the potential for crust formation.

Chisel plowing: Unlike conventional plowing, which inverts the soil, chisel plowing loosens the soil without completely disturbing its structure. This reduces the risk of crust formation while improving water infiltration. Conservation tillage also improves soil moisture retention and reduces the erosive impact of raindrops on the soil surface, both of which help mitigate soil crusting.

3. Improving Water Management Practices

Water management plays a key role in preventing soil crusting by controlling the intensity and distribution of water on the soil surface.

Irrigation management: Applying water in a controlled manner can reduce the likelihood of crust formation. Drip irrigation or low-pressure sprinklers are effective methods for delivering water directly to the root zone without causing significant disturbance to the soil surface. These methods minimize soil particle dislodging and reduce the formation of surface seals.

Avoiding over-irrigation: Over-irrigation can lead to waterlogging, which promotes soil dispersion and crust formation. Proper scheduling of irrigation to match crop needs and soil moisture levels helps prevent crusting.

Rainwater harvesting and contour bunding: In areas prone to heavy rainfall, contour bunding and water harvesting techniques can reduce runoff and improve infiltration. These techniques slow down the movement of water, allowing it to penetrate the soil more effectively and reducing the risk of soil crusting.

4. Enhancing Biological Activity

Biological activity in the soil can improve aggregate stability and prevent soil crusting. The activity of plants, microorganisms, and fauna contributes to soil structure formation and maintenance.

Promoting soil biodiversity: Increasing the diversity of soil organisms, such as earthworms, fungi, and bacteria, can improve soil structure and aggregation. Earthworms, for example, create channels in the soil that improve water infiltration and help bind soil particles together. Fungi, particularly mycorrhizal fungi, contribute to aggregate stability by producing polysaccharides that act as glue to hold particles together.

Encouraging mycorrhizal associations: Mycorrhizal fungi form symbiotic relationships with plant roots, enhancing the stability of soil aggregates. These fungi produce hyphae, which help bind soil particles and improve the soil's physical structure. Encouraging mycorrhizal associations can be achieved by reducing chemical inputs that harm fungi and planting crops that promote their growth.

Cover cropping and root exudates: The root systems of cover crops and other vegetation release organic compounds called exudates, which stimulate microbial activity and promote soil aggregation. The continuous presence of living roots in the soil enhances biological activity and reduces the likelihood of crust formation.

5. Soil Amendments and Conditioning

Certain soil amendments can be applied to improve soil structure and reduce crust formation.

Gypsum (calcium sulfate): In sodic soils, where high sodium levels contribute to dispersion and crusting, gypsum is commonly applied as an amendment. Gypsum replaces sodium ions with calcium ions, improving soil structure and reducing dispersion. The application of gypsum can enhance soil aggregation, improve water infiltration, and reduce crust formation.

Polyacrylamides (PAMs): Polyacrylamides are synthetic polymers that can be applied to the soil to stabilize aggregates and reduce the detachment of soil particles. PAMs are particularly effective in irrigation systems, where they help reduce surface sealing and improve water infiltration by binding soil particles together.

Lime: In acidic soils, liming can improve soil structure by reducing soil acidity and promoting the activity of beneficial soil organisms. Lime also helps to flocculate clay particles, reducing their dispersibility and minimizing crust formation.

6. Maintaining Vegetative Cover

Maintaining vegetative cover is an essential strategy for preventing soil crust formation, as plants protect the soil surface from the direct impact of raindrops and promote better water infiltration.

Cover crops: Growing cover crops, such as legumes, grasses, or brassicas, during the off-season protects the soil from erosion and crust formation. Cover crops provide a protective canopy over the soil, reduce raindrop impact, and contribute organic matter to the soil.

Permanent vegetation: In permanent cropping systems such as orchards or vineyards, maintaining a vegetative cover between rows (such as grass or legumes) reduces the risk of crust formation. The roots of these plants help stabilize the soil and improve its structure over time.

Residue management: Leaving crop residues on the soil surface after harvest helps reduce soil exposure to rain and wind. Residue cover acts as a buffer, protecting the soil from erosion and aggregate breakdown.

Factors Influencing Crust Formation

1. Soil Factors

a) **Soil Organic Matter (SOM):** Soil Organic Matter (SOM) plays a crucial role in maintaining soil structure by acting as a binding agent for both primary and secondary soil particles. It also absorbs water, reducing clay wetting and subsequent breakdown of aggregates. SOM contains three types of binding agents:

Transient agents: These are organic materials that decompose quickly due to microbial activity.

Temporary agents: These include roots and fungal hyphae, particularly vesicular-arbuscular mycorrhizal hyphae, which provide short-term stability.

Persistent agents: These are composed of degraded humic substances associated with amorphous forms of iron, aluminum, and aluminosilicates, offering long-lasting stability. specific fraction of SOM, known as Particulate Organic Matter (POM), plays a pivotal role in enhancing soil aggregation. POM forms organic cores around which clay, silt, and micro-aggregates accumulate, improving soil structure. As a result, the stability of POM is closely linked to the overall stability of soil aggregates.

b) **Soil Texture:** The role of soil texture in crust formation is complex. Studies suggest that as clay content increases, the likelihood of crust formation decreases. Clay particles can help bind soil aggregates together, thereby reducing the disruptive impact of raindrops and protecting the soil from breakdown. Conversely, soils with very low clay content have high infiltration rates but are more prone to pore clogging and weaker aggregate stability. Soils with a higher proportion of clay may have increased aggregate cohesion, minimizing the risk of crusting. This effect, as noted by Kay and Angers (1999), suggests that clay enhances aggregate stability, preventing breakdown upon wetting and thus mitigating crusting.

c) **Clay Mineralogy:** Clay mineralogy strongly influences the dispersion and stability of soils during wetting. The tendency of clay particles to disperse depends on the strength of the attractive forces between them. Certain clay minerals, like kaolinite, exhibit higher stability due to the strong interactions between their positively charged edges and negatively charged planar surfaces. This structure makes kaolinitic soils more stable and less prone to dispersion.

In contrast, smectitic clays tend to be more dispersive due to their morphology and electrical properties. Studies by Wakindiki and Ben-Hur (2002) showed that kaolinitic soils had a significantly higher steady-state infiltration rate (SSIR) compared to smectitic soils, despite similar clay content. While kaolinitic soils form thin, less persistent crusts that break upon drying, smectitic soils tend to form thicker, harder crusts. Overall, soils dominated by kaolinite are more stable and less dispersive than those containing montmorillonite or smectite.

2. Environmental Factors

a) **Vegetation Cover:** Vegetation significantly reduces the strength and formation of soil crusts. Studies have shown that soils under permanent vegetation exhibit much better infiltration and structural stability than exposed, bare soils. Vegetation helps protect the soil surface from direct raindrop impact, thereby reducing the tendency of the soil to form crusts. In areas with natural vegetation, the infiltration rate can be nearly double that of exposed soils, while modulus of rupture and water-dispersible clay content are substantially lower. These changes are attributed to improvements in soil chemistry, which enhance clay flocculation and reduce the likelihood of crust formation (Shainberg & Singer, 1985).

b) **Rainfall:** Rainfall characteristics, particularly raindrop impact and intensity, have a direct influence on soil aggregate breakdown and crusting. The force of raindrops can cause slaking, which initiates the process of physical crust formation by disintegrating soil aggregates. The collapse of micro-aggregates and soil particles that follows alters surface hydraulic processes, such as the steady-state infiltration rate (SSIR) and surface runoff. The combined effects of rainfall intensity and soil characteristics determine how quickly a crust forms and the degree to which it influences water movement and soil erosion (Carmi and Berliner, 2008).

Conclusion

Soil crusting is a significant agricultural problem that reduces water infiltration, increases runoff and erosion, and hinders seedling emergence, ultimately affecting crop productivity. It occurs due to the disintegration of soil aggregates, often caused by heavy rainfall, irrigation, wind, or improper tillage practices, particularly in soils with low organic matter, high silt, or fine sand content. Effective management of soil crusting requires an integrated approach that focuses on improving soil structure, enhancing organic matter content, and minimizing surface disturbance. Strategies such as adding organic amendments (compost, manure, or crop residues), practicing conservation tillage or no-till farming, using cover crops, and applying mulch help maintain soil aggregation and reduce crust formation. Additionally, controlled irrigation methods, chemical amendments (like gypsum for sodic soils), and mechanical techniques (light harrowing or use of soil conditioners) can further mitigate crusting. By adopting these sustainable soil management practices, farmers can enhance soil health, improve water retention, and ensure better crop establishment, leading to long-term agricultural productivity and environmental conservation. Addressing soil crusting is not just about immediate crop benefits but also about preserving soil quality for future generations.

References

Arshad, M. A., & Mermut, A. R. (1988). Micromorphological and physico-chemical characteristics of soil crust types in northwestern Alberta, Canada. *Soil Science Society of America Journal*, *52*(3), 724-729.

Belnap, J. (1999, June). Structure and function of biological soil crusts. In *Sagebrush steppe ecosystems symposium* (pp. 55-62).

Belnap, J., & Lange, O. L. (Eds.). (2013). *Biological soil crusts: structure, function, and management* (Vol. 150). Springer Science & Business Media.

Bolan, N., Nortje, G., Bolan, S., Srivastava, P., Kumar, M., Wang, H., ... & Kirkham, M. B. (2023). Soil Crust: Formation, Influence on Soil Productivity, and Management. In *Soil Constraints and Productivity* (pp. 179-188). CRC Press.

Bresson, L. M., & Boiffin, J. (1990). Morphological characterization of soil crust development stages on an experimental field. *Geoderma*, *47*(3-4), 301-325.

Chamizo, S., Cantón, Y., Miralles, I., & Domingo, F. (2012). Biological soil crust development affects physicochemical characteristics of soil surface in semiarid ecosystems. *Soil Biology and Biochemistry*, *49*, 96-105.

Chamizo, S., Stevens, A., Cantón, Y., Miralles, I., Domingo, F., & Van Wesemael, B. (2012). Discriminating soil crust type, development stage and degree of disturbance in semiarid environments from their spectral characteristics. *European Journal of Soil Science*, *63*(1), 42-53.

Chattopadhyay, S., & Singha, R. Soil Crusting: Causes, Effects and Management Strategies.

Coxson, D. S. (2002). Biological soil crusts: structure, function, and management. *The Bryologist*, *105*(3), 500-501.

Eldridge, D. (2000). Ecology and management of biological soil crusts: recent developments and future challenges. *The Bryologist*, *103*(4), 742-747.

Evans, D. D., & Buol, S. W. (1968). Micromorphological study of soil crusts. *Soil Science Society of America Journal*, *32*(1), 19-22.

Gicheru, P., Gachene, C., Mbuvi, J., & Mare, E. (2004). Effects of soil management practices and tillage systems on surface soil water conservation and crust formation on a sandy loam in semi-arid Kenya. *Soil and Tillage Research*, *75*(2), 173-184.

Kakeh, J., Gorji, M., Mohammadi, M. H., Asadi, H., Khormali, F., Sohrabi, M., & Cerdà, A. (2020). Biological soil crusts determine soil properties and salt dynamics under arid climatic condition in Qara Qir, Iran. *Science of the total environment*, *732*, 139168.

Niu, J., Yang, K., Tang, Z., & Wang, Y. (2017). Relationships between soil crust development and soil properties in the desert region of North China. *Sustainability*, *9*(5), 725.

Paglial, M. (2005). Soil crusting. *College of Soil Physics ICIP*.

Pagliai, M., Lamarca, M., & Lucamante, G. (1983). Micromorphometric and micromorphological investigations of a clay loam soil in viticulture under zero and conventional tillage. *Journal of Soil Science*, *34*(2), 391-403.

Rivera-Aguilar, V., Godínez-Alvarez, H., Moreno-Torres, R., & Rodríguez-Zaragoza, S. (2009). Soil physico-chemical properties affecting the distribution of biological soil crusts along an environmental transect at Zapotitlán drylands, Mexico. *Journal of Arid Environments*, *73*(11), 1023-1028.

Tackett, J. L., & Pearson, R. W. (1965). Some characteristics of soil crusts formed by simulated rainfall. *Soil Science*, *99*(6), 407-413.

Thomas, A. D., & Dougill, A. J. (2006). Distribution and characteristics of cyanobacterial soil crusts in the Molopo Basin, South Africa. *Journal of Arid Environments*, *64*(2), 270-283.

Valentin, C. (2005). Surface crusts of semi-arid sandy soils: types, functions and management. *FAO proceedings Management of Tropical sandy soils for sustainable agriculture. 27th November to 2nd December*.

Xu, H., Zhu, X., & Mi, M. (2025). Progress and Prospects of Research on Physical Soil Crust. *Soil Systems*, *9*(1), 23.

Zhao, H. L., Guo, Y. R., Zhou, R. L., & Drake, S. (2010). Biological soil crust and surface soil properties in different vegetation types of Horqin Sand Land, China. *Catena*, *82*(2), 70-76.

9

Soil Erosion

Introduction

Soil, the outermost layer of the Earth's crust, plays a vital role in supporting plant life. It consists of a mixture of mineral and organic materials, with its depth varying from negligible to several meters across different regions. Subjected to continuous atmospheric influences, soil experiences erosion, primarily driven by the movement of wind and water. Erosion, originating from the Latin word "erodere," meaning "to eat away" or "to excavate," involves the detachment, transportation, and deposition of soil particles.

The erosion process encompasses two main phases: the detachment of individual soil particles from the soil mass and their subsequent transportation by various agents such as water, wind, ice, or gravity, followed by eventual deposition. Finer soil particles are more susceptible to erosion compared to coarser ones, leading to the formation of sediment, the ultimate product of soil erosion. Sediment, which refers to fragmented material transported or deposited by natural agents, illustrates the sedimentation process, reflecting the sequence of the sediment cycle.

Detachment involves the dislodgment of soil particles from the soil mass by erosive agents such as impacting raindrops and runoff water. Transportation entails the entrainment and movement of these detached soil particles (sediment) from their original location, often through stream systems, towards eventual deposition sites, which may include the base of slopes, reservoirs, or floodplains.

Soil erosion presents a significant threat to human well-being by diminishing the productivity of agricultural land through the removal and washing away of essential plant nutrients and organic matter. The widespread distribution of sediment load globally underscores the pervasive impact of erosion on natural landscapes and ecosystems.

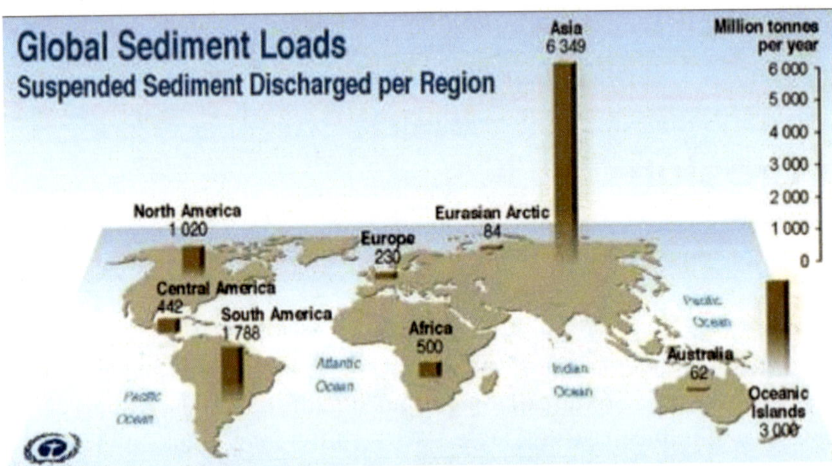

Figure 1. Asia has the highest suspended sediment discharge globally, primarily due to its substantial monsoon rainfall. (*Source:* Peter, 1983).

Challenges Arising from Soil Erosion (Telles *et al.*, 2011)

The harmonious balance of ecosystems, encompassing soil, water, and plant environments, is vital for the well-being and survival of humanity. Yet, across various regions worldwide, including certain areas in India, ecosystems have suffered from past overexploitation, leading to disturbances. These disruptions manifest in various adverse effects, including soil surface degradation and the frequent occurrence of extreme floods.

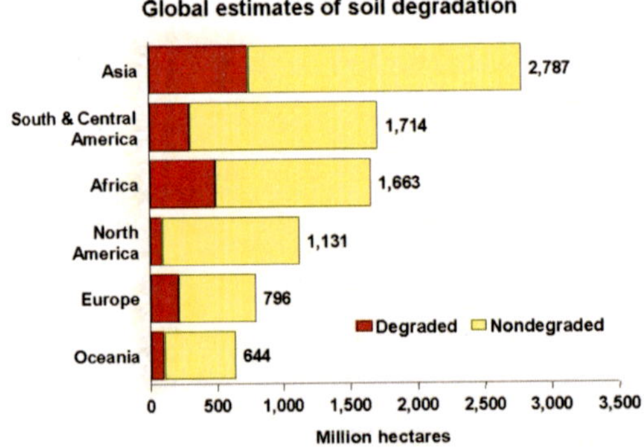

Figure 2. The scale of land degradation across continents (*Source:* Scherr, 2019)

Large tracts of land have irreversibly transformed into barren surfaces due to swift soil erosion induced by a variety of factors. This degradation of land has further worsened the pollution of natural water sources. The accumulation of eroded soil from higher elevations into lower river reaches has caused in aggradation, expanding floodplain areas, diminishing clearance under bridges and culverts, and leading to sedimentation in reservoirs. The scale of land degradation at a continental level is depicted in Figure 2. The major challenges arising from sedimentation-related land degradation are succinctly described below.

2.1 Water and wind erosion

Approximately 69 million hectares of the 144.12 million hectares impacted by wind and water erosion are categorized as critical, requiring immediate intervention. Wind erosion predominantly impacts states like Gujarat, Rajasthan and Haryana. The intensity of wind erosion shows an inverse relationship with rainfall levels, with regions experiencing lower rainfall exhibiting higher rates of wind erosion.

2.2 Ravines and Gullies

Approximately 4 million hectares across 12 states in the country are afflicted by gullies and ravines. Ravines predominantly occur in Madhya Pradesh, Gujarat, Uttar Pradesh, and Rajasthan, while gullies are prevalent in the plateau regions of Eastern India, foothills of the Himalayas, and Deccan Plateau areas.

2.3 Torrents and Riverine Lands

The issue of riverine and torrents affects approximately 2.73 million hectares in the country. Torrents, natural streams prone to frequent alterations in their course and sudden, heavy debris-laden flash floods, pose significant threats to life and property. The infertile material or debris carried by torrents is often deposited onto fertile plains, resulting in irreversible damage to the land.

2.4 Water logging

Waterlogging arises from either surface flooding or a rise in the water table. Approximately 8.53 million hectares are affected by waterlogging, with surface flooding being prevalent in states such as Bihar, Assam, Odisha, Kerala, West Bengal, Uttar Pradesh, Punjab, Andhra Pradesh, and Haryana.

2.5 Shifting Cultivation

Shifting cultivation, also known as "jhuming," is a conventional agricultural method where crops are grown on hill slopes using slash-and-burn methods. This practice involves selecting suitable sites on hills, clearing forests by burning and cutting vegetation, cultivating the land for a few years, and then moving to a new site. Due to population growth pressures, the jhum cycle has progressively decreased from 20-30 years to 3-6 years. This challenge is particularly prominent in the northeastern region, along with in the states of Odisha and Andhra Pradesh.

2.6 Saline soils, which include coastal regions

Saline soils are widespread in both inland and coastal regions of the country, encompassing approximately 5.5 million hectares of land. These areas include the semi-arid and arid regions of Rajasthan and Gujarat, the black soil belt, and coastal areas. This problem presents a major challenge to agricultural areas, rendering fertile soil unproductive and causing brackish groundwater in states like Odisha, Tamil Nadu, West Bengal, Maharashtra, Kerala, Karnataka, Gujarat, and Andhra Pradesh, as well as the Union Territories of Pondicherry, Goa, and Daman and Diu.

2.7 Droughts and Floods

In India, among the major and medium rivers categorized in both the Himalayan and non-Himalayan regions, 18 rivers are prone to flooding, covering a combined area of 150 million hectares. Recent occurrences of flash floods have produced substantial damage, extending even to the arid regions of Gujarat and Rajasthan.

3. Significance of Soil Conservation (Bilotta *et al.*, 2012)

In India, out of a total geographical area spanning 329 million hectares, approximately 150 million hectares are susceptible to either water or wind erosion. Currently, around 140 million hectares are cultivated. Moreover, 40 million hectares face the risk of flooding, while approximately 4 million hectares have been lost to ravines and gullies. Overall, an estimated 175 million hectares, constituting 53.3% of the total geographical area, are affected by various soil and land degradation issues, including saline-alkali soils, waterlogged areas, ravine and gullied lands, shifting cultivation areas, and desertification.

Looking forward to the year 2100 A.D., India's projected population is set to reach two billion, while food grain production has remained stagnant at 211 million tons for the past five years. The per capita cropped area has steadily declined from 0.33 hectares per capita in 1950 to 0.15 hectares per capita by 2000. This highlights the critical importance of managing the limited land resources meticulously to ensure the sustenance of the growing population.

To safeguard soil and water resources in India, several recommendations are proposed:

- Upholding natural cover, particularly on susceptible terrains like steep slopes during dry spells or heavy rainfall periods, to avert erosion of exposed soil.

- Implementing selective harvesting practices in forests, such as alternating tree cuts, and Utilizing seasonally dry or wet areas for pasture instead of converting them into arable land can be a more sustainable approach.
- Implementing crop rotation in areas of intensive cultivation to thwart soil depletion and adopting contour ploughing methods in vulnerable zones.
- Prudently managing irrigation practices to prevent soil salinity build-up and implementing measures to curb overgrazing by livestock.

Limiting the construction of highways and urban development to regions with lower agricultural potential and ensuring land restoration post-extractive industry operations can help mitigate environmental degradation. These conservation strategies are imperative for securing sustainable land and water management for future generations in India.

4. The historical narrative of soil erosion and soil conservation initiatives in India, (Mandal & Giri, 2021)

The historical evolution of soil erosion and conservation initiatives in India reflects a journey of recognizing and addressing the adverse impacts of soil degradation driven by human activities and environmental factors. From the pre-independence era, when pioneers like Sir Dietrich Brands advocated for tree planting to combat erosion, to the post-independence period marked by the establishment of dedicated soil conservation boards and research centers, the trajectory of soil conservation in India has been one of concerted efforts and evolving strategies.

During the early decades post-independence, systematic soil conservation initiatives gained momentum, with watershed management projects and contour bunding becoming key components of soil conservation efforts. The establishment of the Central Soil Conservation Board (CSCB) and the subsequent focus on research, training, and demonstration centers underscored the government's commitment to addressing soil erosion comprehensively.

As India progressed through successive five-year plans, the scope and scale of soil conservation programs expanded, encompassing diverse ecosystems and regions. Programs like the National Watershed Development Programme for Rainfed Areas (NWDPRA) and the Desert Development Programme (DDP) exemplify the government's multifaceted strategy towards soil and water resource management.

In the Tenth Five Year Plan, there was a notable pivot towards natural resource management, placing considerable emphasis on groundwater recharge, rainwater harvesting, and watershed development. Fully funded initiatives like the Western Ghats Development Programme (WGDP) aimed at erosion-prone regions, implementing a range of soil conservation measures to alleviate environmental degradation.

In the Eleventh Five Year Plan, watershed development projects received renewed attention, with a focus on enhancing water resource availability and promoting efficient utilization. Water management responsibilities were decentralized among various government departments, reflecting a coordinated approach to address the complex challenges of soil erosion and conservation.

Overall, India's journey in soil conservation has been characterized by a blend of traditional wisdom, scientific innovation, and institutional collaboration. Despite the persistent challenges posed by soil erosion, the nation's commitment to sustainable land and water management remains steadfast, ensuring the preservation of precious natural resources for future generations.

5. Causes of Soil Erosion (Balasubramanian, 2017)

Soil erosion cannot be ascribed to a single unique cause; rather, it results from a combination of factors, some influenced by natural forces and others by human activities. The primary causes of soil erosion include:Top of Form

- Destruction of Natural Protective Cover:
- Unregulated deforestation
- Overgrazing of vegetation
- Incidents of forest fires
- Improper Land Use:
- Leaving land barren, exposing it to rain and wind erosion
- Cultivating crops that exacerbate soil erosion
- Depletion of organic matter and nutrients through unsustainable cropping practices
- Cultivation on slopes, increasing susceptibility to erosionTop of Form
- Faulty methods of irrigation

6. Types of Soil Erosion (Telkar *et al.*, 2015)

6.1 Based on the Origin

Soil erosion can be generally classified into two types

- Geologic erosion
- Accelerated erosion

6.1.1 Geological Erosion

At natural undisturbed conditions, a harmonious equilibrium exists between the vegetative cover and climate of an area, acting as a protective layer for the soil. Vegetative covers such as trees and forests play a essential role in slowing down soil transportation and acting as a barrier against extreme erosion. Even in the presence of natural cover, a degree of erosion, termed geologic erosion, persists. This process is gradual and is offset by soil formation through natural weathering processes. Geologic erosion typically has minimal impact on agricultural lands.

6.1.2 Accelerated Erosion

When land undergoes cultivation, it disturbs the harmonious equilibrium between soil, climate. and vegetation cover. Under these conditions, natural mechanisms result in the faster removal of surface soil compared to its replenishment through soil formation. This process, termed accelerated erosion, exceeds the rates observed in geological erosion. Accelerated erosion significantly diminishes soil fertility, particularly in agricultural regions.

6.2 Based on the Erosion Agents

Soil erosion is classified into various types depending on the agents that initiate the erosion process. The four key types of soil erosion include:

6.2.1 Water Erosion

Water erosion is a widespread phenomenon across various regions globally, primarily driven by the action of running water as the dominant agent of soil erosion. This includes rivers eroding river basins, rainfall eroding various landforms, and sea waves eroding coastal areas. The process of water erosion entails the detachment and transportation of soil particles from higher elevations to lower-lying areas. Moreover, water erosion can be classified based on the diverse actions of water responsible for erosion, such as:

- Raindrop erosion
- Sheet erosion
- Rill erosion
- Gully erosion
- Stream bank erosion
- Slip erosion

6.2.2 Wind Erosion

Wind erosion predominantly occurs in arid regions where powerful winds interact with diverse landforms, causing them to erode and dislodge soil particles. These particles are subsequently lifted and carried in the direction of the prevailing wind. Iconic examples of wind erosion include mushroom rock formations and sand dunes, commonly observed in desert landscapes.

6.2.3 Glacial Erosion

Glacial erosion, or ice erosion, is common in cold regions at high altitudes. Soil that comes into contact with large moving glaciers sticks to the base of the glacier. As the glacier moves, it carries the soil with it, and when it melts, the soil is deposited along the path of the moving ice chunks.

6.2.4 Gravitational Erosion

While less prevalent compared to water erosion, gravitational erosion poses substantial risks to both natural and man-made structures. Gravitational erosion affects both natural landscapes and man-made structures, involving the mass movement of soil driven by gravitational forces. Landslides and slumps exemplify this phenomenon, occurring swiftly, while soil creep unfolds gradually over extended periods.

7. Soil Erosion Agents (Adhikary, 2020)

Soil erosion encompasses The separation of soil from its original location and its subsequent relocation to another location. While water predominantly drives this process, wind and glaciers also play roles in soil erosion across different areas. Water, in the forms of rain, floods, and runoff, significantly impacts soil. Soil, comprised of sand, silt, and clay, is vulnerable to erosion when rainfall occurs over mountainous terrain or bare soil, as water dislodges soil particles and carries away clay and silt with flowing water. Likewise, strong winds during storms have the capacity to lift the upper layer of soil, causing erosion. Animal and human activities also play a role in soil erosion. Vegetation serves as a natural soil cover, but continuous grazing by animals can remove vegetation, leaving bare lands susceptible to erosion. Human activities like deforestation, intensified agriculture, and land clearance for various purposes exacerbate soil erosion. The agents accountable for soil erosion can be classified and outlined as depicted in Figure 3.

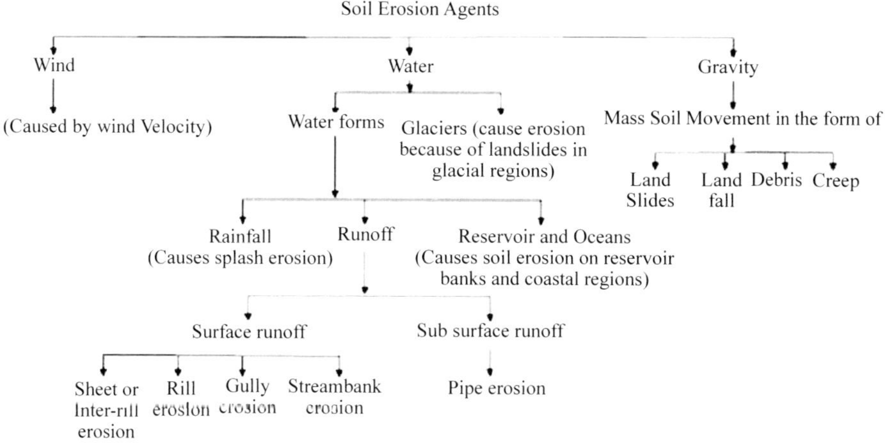

Figure 3. Agents of soil erosion (*Source:* Adhikary, 2020)

8. Factors Affecting Soil Erosion (El-Swaify, 1997)

Soil erosion involves the processes of soil particle detachment from the soil mass, followed by their transportation and eventual deposition. In India, significant factors contributing to soil erosion include extensive deforestation, overgrazing, and improper agricultural practices. The complexity of soil erosion arises from the interaction of multiple interconnected factors affecting its rate. These factors include:

8.1 Climatic Factor

Climatic factors, including the amount, intensity, and frequency of rainfall, exert considerable influence on erosion processes. In instances of recurrent or continuous rainfall, soil moisture levels rise, resulting in saturated field conditions. This amplifies the alteration of rainfall into runoff, promoting soil separation and transportation, thereby accelerating erosion rates.

8.2 Temperature

Frozen soil commonly displays strong resistance against erosion. Yet, swift thawing of the soil surface caused by warm rains can precipitate notable erosion. Temperature also determines the form of precipitation. While falling snow generally does not contribute to erosion, extensive snowmelt during spring thaw can result in substantial runoff damage. Additionally, temperature plays a role in the accumulation of organic matter on the ground surface and its integration into the topsoil layer. Top of Form Warmer regions often exhibit thinner organic soil cover. This surface organic matter serves as a protective barrier against the impact of rainfall and aids in rainfall infiltration, thereby reducing runoff. Additionally, organic matter within the soil enhances soil permeability, promoting increased percolation and reduced runoff.

8.3 Topographical Factors

Topographical factors such as slope length, steepness, and roughness significantly influence erodibility. Extended slopes typically exacerbate erosion potential, particularly at their base wherever runoff velocity peaks and accumulates. The slope's steepness, along with surface roughness and rainfall intensity, collectively determines the speed of runoff down a slope. Greater slope angles expedite water flow, heightening erosion and sedimentation risks. Slope inclination amplifies erosion by accelerating water velocity. Minor slope discrepancies can yield significant repercussions. In accordance with hydraulic principles, a fourfold rise in slope doubles water velocity, leading to a quadruple increase in erosive force and a 32-fold surge in sediment-carrying capacity.

8.4 Soil

The erodibility of soil is significantly influenced by its physical characteristics, encompassing texture, cohesion and structure. Texture pertains to the size or combination of sizes of individual soil particles, broadly classified as clay, silt, and sand. Soils rich in silt-sized particles are particularly susceptible to erosion by wind and water, while those with clay or sand-sized particles are less vulnerable. Structure, on the other hand, refers to the organization of soil particles into larger clumps and pore spaces, affecting water absorption and resistance to erosion. Cohesion, on the other hand, represents the binding force between soil particles, directly affecting soil structure. Clay soils exhibit high cohesion, leading to a doughy consistency when moist, whereas sand soils display the lowest cohesion.

8.5 Vegetation

Vegetation stands out as a crucial physical factor affecting soil erosion significantly. A dense vegetative cover acts as a shield against the erosive impact of raindrops and promotes cohesion among soil particles, thereby enhancing resistance to runoff. Moreover, vegetation contributes organic matter, moderates runoff, and acts as a natural filter for sediment. Especially on graded slopes, the density and health of vegetative cover dictate the extent to which erosion is prevented or mitigated. A lush, dense vegetative canopy stands as one of the most potent defences against soil erosion.

8.6 Biological Factors of Soil Erosion

Biological factors impacting soil erosion encompass practices such as faulty cultivation methods and overgrazing by animals. These factors can be broadly categorized into the following three groups:

- Energy factors
- Resistance factors
- Protection factors

8.6.1 Energy Factors

These factors encompass influences on the potential of rainfall, runoff, and wind to induce erosion, collectively known as erosivity. Additionally, measures such as constructing terraces and bunds to shorten or reduce the slope length/degree in water-eroded areas, along with establishing shelter belts or windbreaks in wind-eroded areas, directly diminish the erosive power of these agents.

8.6.2 Resistance Factors

These factors, termed erodibility factors, depend on both the mechanical and chemical attributes of the soil. Conditions promoting water infiltration into the soil decrease runoff and lower erodibility, while practices leading to soil pulverization increase it. As a result, cultivation may reduce the erodibility of clay soils but increase it for sandy soils.

8.6.3 Protection Factors

This primarily underscores the factors related to vegetation cover. Vegetation acts as a protective barrier against erosion by intercepting rainfall and reducing the speed of runoff and wind. The effectiveness of different vegetation covers varies widely. Hence, it is crucial to evaluate soil erosion rates across various land uses, vegetation covers and slope gradients to determine suitable land management practices for controlling erosion rates. Soil loss, quantified as the amount of soil moved beyond a specific location, is commonly expressed as mass or volume per unit area per unit time.

9. Mechanics of Soil Erosion (Telkar *et al.*, 2015)

Soil erosion initiates with the detachment (seperation) of soil particles prompted by rainfall. These dislodged particles are then carried by erosion agents to different areas, where they settle, perpetuating the soil erosion cycle. Figure 4 shows various soil erosion processes.

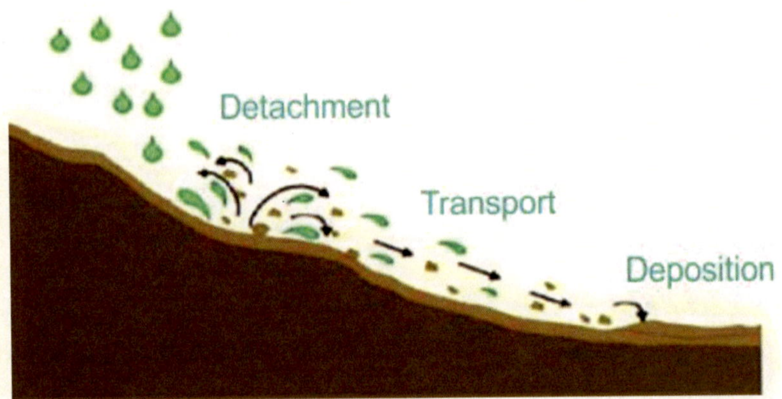

Figure 4. Process of soil erosion (*Source:* Zafirah *et al.*, 2017)

10. Erosion Control Methods (Ahmad *et al.*, 2020)

10.1 Planting Vegetation

This approach involves cultivating crops with deep roots that can effectively anchor the soil, particularly in erosion-prone areas like streams, slopes, and riverbanks. Vegetative barriers, formed by densely packed thick stems, impede the flow of water, allowing it to pass through gently without causing erosion. Utilizing deep-rooted native plants such as wildflowers, woody perennials, and native prairie grasses is highly recommended for erosion control measures.

10.2 Contour Farming

Managing and cultivating slopes can pose challenges, often leading to rapid soil erosion. Yet, contour farming offers a promising solution. By planting crops perpendicular to the slope along contour lines, farmers can mitigate erosion effectively. This method conserves rainfall and minimizes surface erosion, utilizing crop rows, vehicle tracks, and furrows to retain rainwater. Consequently, these practices serve as reservoirs for rainfall, enhancing soil retention and agricultural sustainability.

10.3 Geotextiles

Geotextile walls offer a sustainable solution to erosion control, commonly used for stabilization purposes. These structures consist of porous fabrics that, when combined with soil, serve various functions such as separation, filtration, reinforcement, protection, and drainage. By allowing water to drain away while retaining soil particles, geotextile walls effectively prevent runoff and sedimentation. Additionally, they contribute to reducing overall pollution levels. Whether in the form of non-woven or woven geotextiles, these materials can be applied to steep hillsides to create a protective barrier against erosion.

10.4 Applying Mulches

In this approach, mulch components are applied to cover bare soil, safeguarding it from erosion. Mulching is commonly used during the early stages of seedling or bush growth to prevent soil loss. Additionally, mulch helps retain moisture and regulates soil temperature, minimizing fluctuations. Wood mulch is suitable for gardens and landscapes, while organic mulches can nourish and shield plants, particularly during spring and fall.

10.5 Reforestation

Restoring damaged ecosystems and conserving present ones are crucial for effective soil erosion management. Recent research indicates that appropriately planted and maintained trees have the potential to decrease erosion by as much as 75%. Furthermore, the depletion of forest cover heightens the susceptibility to soil erosion and earth flow, as it disrupts the protective forest canopy and severs the intricate network of interlinked roots in the subsoil.

11. Conclusion

In conclusion, soil erosion presents a serious threat to our environment and agricultural productivity. The historical evolution of soil conservation programs underscores the ongoing efforts to address this challenge. Grasping the underlying causes, types, and mechanisms of soil erosion is essential for the successful implementation of control measures. By considering factors such as climate, topography, vegetation cover, and soil properties, sustainable erosion control methods can be devised. Through collaborative efforts and innovative approaches, we can mitigate the impacts of soil erosion and Guarding our natural resources for the well-being of future generations.

References

Adhikary, R. (2020). Causes and Effect of Soil Erosion and its Preventive Measures. Advanced Agriculture by S. Maitra and B. Pramanick (Editors). New Delhi Publishers, New Delhi, 376-387.

Ahmad, N. S. B. N., Mustafa, F. B., & Didams, G. (2020). A systematic review of soil erosion control practices on the agricultural land in Asia. International Soil and Water Conservation Research, 8(2), 103-115.

Balasubramanian, A. (2017). Soil erosion–causes and effects. Centre for Advanced Studies in Earth Science, University of Mysore, Mysore.

Bhat, S. A., Dar, M. U. D., & Meena, R. S. (2019). Soil erosion and management strategies. Sustainable management of soil and environment, 73-122.

Bilotta, G. S., Grove, M., & Mudd, S. M. (2012). Assessing the significance of soil erosion. Transactions of the institute of British Geographers, 37(3), 342-345.

Colazo, J. C., Carfagno, P., Gvozdenovich, J., & Buschiazzo, D. (2019). Soil erosion. The soils of Argentina, 239-250.

El-Swaify, S. A. (1997). Factors affecting soil erosion hazards and conservation needs for tropical steeplands. Soil technology, 11(1), 3-16.

Gleick, P. H. (1993). Water in crisis (Vol. 100). New York: Oxford University Press.

Kairis, O., Karavitis, C., Kounalaki, A., Salvati, L., & Kosmas, C. (2013). The effect of land management practices on soil erosion and land desertification in an olive grove. Soil Use and Management, 29(4), 597-606.

Lal, R. (1990). Soil erosion in the tropics: principles and management (pp. x+-580).

Mandal, D., & Giri, N. (2021). Soil erosion and policy initiatives in India. Current science, 1007-1012.

Misir, N., Misir, M., Karahalil, U., & Yavuz, H. (2007). Characterization of soil erosion and its implication to forest management. Journal of Environmental Biology, 28(2), 185-191.

Morgan, R. P. C. (2009). Soil erosion and conservation. John Wiley & Sons.

Nikkami, D., Elektorowicz, M., & Mehuys, G. R. (2002). Optimizing the management of soil erosion. Water Quality Research Journal, 37(3), 577-586.

Novara, A., Gristina, L., Saladino, S. S., Santoro, A., & Cerdà, A. (2011). Soil erosion assessment on tillage and alternative soil managements in a Sicilian vineyard. Soil and Tillage Research, 117, 140-147.

Panagos, P., Borrelli, P., Meusburger, K., Alewell, C., Lugato, E., & Montanarella, L. (2015). Estimating the soil erosion cover-management factor at the European scale. Land use policy, 48, 38-50.

Pennock, D. (2019). Soil erosion: The greatest challenge for sustainable soil management. FAO;.

Renschler, C. S., & Harbor, J. (2002). Soil erosion assessment tools from point to regional scales—the role of geomorphologists in land management research and implementation. Geomorphology, 47(2-4), 189-209.

Scherr, S. J. (2019). The future food security and economic consequences of soil degradation in the developing world. Response to land degradation, 155-170.

Telkar, S. G., Solanki, S. P. S., Dey, J. K., & Kant, K. (2015). Soil erosion: Types and their mechanism. International Journal of Economic Plants, 2(4), 178-180.

Telles, T. S., Guimarães, M. D. F., & Dechen, S. C. F. (2011). The costs of soil erosion. Revista Brasileira de Ciencia do solo, 35, 287-298.

Vanwalleghem, T., Gómez, J. A., Amate, J. I., De Molina, M. G., Vanderlinden, K., Guzmán, G., ... & Giráldez, J. V. (2017). Impact of historical land use and soil management change on soil erosion and agricultural sustainability during the Anthropocene. Anthropocene, 17, 13-29.

Zafirah, N., Nurin, N. A., Samsurijan, M. S., Zuknik, M. H., Rafatullah, M., & Syakir, M. I. (2017). Sustainable ecosystem services framework for tropical catchment management: A review. Sustainability, 9(4), 546.

10

Quality of Irrigation Water

Introduction

Water is one of the most vital natural resources for sustaining life on Earth. However, its availability is not uniform across the planet, and the growing demand for freshwater in the face of climate change, population growth, and industrial development has raised serious concerns about the future of water resources. In 2024, the global water landscape presents both challenges and opportunities, shaped by shifting environmental patterns, technological advances, and policy responses to water scarcity.

Water resources refer to the natural sources of water that can be utilized for various purposes, including water supply. Approximately 97% of the Earth's water is saline, found primarily in oceans, while only about 2.5% is fresh water, but only a small portion of this is accessible. The majority of freshwater is locked in glaciers, ice caps, and deep underground aquifers. Less than 1% of Earth's freshwater are readily available in rivers, lakes, and accessible groundwater for human use. Of this small fraction, nearly two-thirds is locked in glaciers and polar ice caps. The remaining unfrozen fresh water exists predominantly as groundwater, with only a small percentage available as surface water in rivers, lakes, and atmospheric moisture. The primary natural sources of freshwater include surface water, underflow in riverbeds, groundwater, and frozen water in glaciers. In addition to these, artificial sources such as treated wastewater (reclaimed water) and desalinated seawater are increasingly being used to supplement the global freshwater supply, particularly in regions facing water scarcity.

As agriculture accounts for approximately 70% of global freshwater withdrawals, the management and quality of irrigation water are vital for sustaining food production and maintaining the health of ecosystems. Water is essential not only for basic plant functions, such as photosynthesis and nutrient absorption, but also for controlling temperature and supporting soil structure. In regions where rainfall is insufficient or irregular, irrigation provides a reliable source of water to meet the needs of crops throughout their growth cycle.

Sources of Irrigation Water

The sources of irrigation water vary based on geographic location, availability, and specific agricultural needs. Common sources include:

Surface water: Rivers, lakes, reservoirs, and streams are often tapped for irrigation. The quality of surface water can fluctuate depending on seasonal changes, pollution levels, and the management of upstream activities.

Groundwater: Wells and aquifers provide groundwater for irrigation. Groundwater quality is generally more stable but may contain dissolved minerals or salts that can affect crop growth over time.

Recycled or treated wastewater: In water-scarce regions, treated wastewater is increasingly used for irrigation. It can provide a reliable supply of nutrients, but strict monitoring is required to prevent the buildup of harmful chemicals, pathogens, or salts.

Rainwater: Collected rainwater is sometimes used for irrigation, especially in water-conserving systems. Its quality is typically high, though contamination from collection surfaces can be a concern.

Quality Parameters of Irrigation Water

The quality of irrigation water is determined by several physical, chemical, and biological parameters. These parameters impact plant growth, soil properties, and the long-term viability of agricultural systems. Agricultural crops receive moisture for their growth and development mostly from two sources: (i) rain water and (ii) irrigation water. Various criteria for the evaluation of irrigation water with permissible limits for crop growth are being discussed below:

1. Salinity hazard or total soluble salt concentration or EC

The concentration of soluble salts in irrigation water is typically measured by electrical conductivity (EC) and expressed in units of dSm^{-1}.

Water class	Salt concentration ($g\ L^{-1}$)	EC (dSm^{-1})	Remarks
Low salinity (C1)	< 0.16	0.0 – 0.25	Can be used safely
Medium salinity (C2)	0.16 – 0.50	0.25 – 0.75	Can be used with moderately leaching
High salinity (C3)	0.50 – 1.50	0.75 – 2.25	Can be used for irrigation purposes with some management practices
Very high salinity (C4)	1.50 – 3.00	2.25 – 5.00	Cannot be used for irrigation purposes

Impact on Agriculture: High salinity in irrigation water can lead to the buildup of salts in the root zone, reducing the availability of water to plants and causing "osmotic stress." Over time, salinization can harm crops, reduce yields, and degrade soil structure.

2. Salt Index

The salt index is a critical parameter used to assess the suitability of irrigation water that contains salts, particularly when the concentration of sodium chloride exceeds safe levels. This index helps predict potential sodium hazards that could impact soil and crop health. It is calculated based on the concentrations of sodium ions (Na^+), calcium ions (Ca^{2+}), and calcium carbonate ($CaCO_3$) present in the irrigation water. All salt concentrations are typically expressed in parts per million (ppm). By evaluating the salt index, farmers and water managers can make informed decisions to prevent soil salinization and mitigate potential negative effects on plant growth and soil structure.

Salt Index = (Total Na – 24.5) – [(Total Ca – Ca in CaCO3) x 4.85]

The salt index is used to assess water quality for irrigation. A negative value indicates good quality water that is suitable for irrigation, while a positive value signifies water that is unsuitable for irrigation due to high salinity.

3. Sodium hazard (SAR)

High concentrations of sodium ions (Na^+) in irrigation water are undesirable because they can adversely affect soil structure. Sodium ions tend to adsorb onto the cation exchange sites in the soil, which can lead to the breakdown of soil aggregates—a process known as de-flocculation. This disruption of soil structure can cause pore spaces to become sealed, impeding water infiltration and drainage.

The extent to which sodium ions dominate the cation exchange sites, displacing other essential cations like calcium (Ca^{2+}) and magnesium (Mg^{2+}), is quantified by the Sodium Adsorption Ratio (SAR). SAR is calculated as the ratio of sodium ion concentration to the combined concentration of calcium and magnesium ions in the irrigation water. This ratio provides an estimate of the potential for sodium-induced soil degradation and helps in assessing the suitability of irrigation water for maintaining healthy soil structure.

Sodium Adsorption Ratio (SAR) is a measure used to assess the relative proportion of sodium (Na^+) to calcium (Ca^{2+}) and magnesium (Mg^{2+}) in water. It is calculated as the ratio of the concentration of sodium ions to the combined concentration of calcium and magnesium ions, with all concentrations expressed in mill equivalents per liter (meq/L). SAR is an important parameter for evaluating the potential impact of irrigation water on soil structure and fertility.

$$SAR = \frac{(Na^{+})}{\sqrt{\frac{(Ca^{2+}) + (Na^{+})}{2}}}$$

Water class	SAR value	Remarks
S1 – Low Na^{+}	< 10	Any type of crops grown and water used
S2 – Medium Na^{+}	10 – 18	Drainage water used for sandy soil
S3 – High Na^{+}	18 – 26	Sensitive crops are not take
S4 – Very high Na^{+}	> 26	This water is not used for crops.

4. Bicarbonate hazard (RSC)

The bicarbonate ion (HCO_3^-) is a crucial component in irrigation water quality. The concentration of bicarbonates is particularly important when assessing water for irrigation because it influences soil chemistry and plant health. Residual Sodium Carbonate (RSC) is a key measure used to evaluate the suitability of irrigation water. RSC is calculated by assessing the balance between bicarbonate and carbonate ions relative to calcium and magnesium ions. High RSC values can indicate a risk of soil sodicity, which can negatively impact soil structure and crop growth.

RSC (me L^{-1}) = (CO32- +HCO3-) – (Ca^{2+} + Mg^{2+})

Water class	RSC value	Remarks
Low RSC	< 1.25	Can be used safely
Medium RSC	1.25 – 2.50	Can be used with certain management
High RSC	> 2.50	Unsuitable for irrigation purposes

5. Boron Concentration

Boron is a crucial micronutrient necessary for the normal growth and development of plants, though it is required in very small quantities. However, high concentrations of boron in irrigation water can be toxic to plants and detrimental to crop health. Therefore, it is essential to monitor and assess boron levels when evaluating water quality for irrigation.

Permissible limits for boron in irrigation water are established to ensure safe and effective use. Exceeding these limits can lead to boron toxicity, which may cause leaf damage, reduced crop yields, and other negative impacts on plant health. The following are the recommended permissible limits for boron in irrigation water:

Water class	Boron concentration (ppm)			Remarks
	Sensitive crops	Semi tolerance crops	Tolerance crops	
Very Low	< 0.33	< 0.67	< 1.00	Can be used safely
Low	0.33 – 0.67	0.67 – 1.33	1.00 – 2.00	Can be used with management
Medium	0.67 – 1.00	1.33 – 2.00	2.00 – 3.00	Unsuitable for irrigation purposes
High	1.00 – 1.25	2.00 – 2.50	3.00 – 3.75	
Very High	> 1.25	> 2.50	> 3.75	

Boron Sensitivity in Crops

Boron-Sensitive Crops

These crops are highly susceptible to boron toxicity and typically exhibit poor growth or yield reductions when exposed to elevated levels of boron. Sensitive crops include:

Apple, Grape, Cherry, Peach, Orange and Lemon

Boron Semi-Tolerant Crops

These crops can tolerate moderate levels of boron but may still show some negative effects if boron concentrations are high. Semi-tolerant crops include:

Sunflower, Potato, Cotton, Tomato, Radish, Field Pea, Barley, Wheat, Corn, Rice, Oat and Sweet Potato

Boron-Tolerant Crops

These crops are more resilient to higher levels of boron and generally exhibit less sensitivity to boron toxicity. Tolerant crops include:

Date Palm, Sugar Beet, Garden Beet, Alfalfa, Onion, Turnip, Cabbage, Lettuce, Carrot and Cauliflower

6. Chloride Concentration

The concentration of chloride ions in irrigation water tends to rise alongside electrical conductivity (EC) and sodium ions, making chloride ions more prevalent in high-salinity water. Unlike sodium ions, chloride ions do not significantly affect the physical properties of the soil nor are they adsorbed onto soil particles. As a result, chloride is generally not included in modern soil classification systems. However, it remains a factor in some regional water classification schemes due to its influence on water quality and its potential effects on specific agricultural practices.

$$\text{Chloride concentration } \left(\frac{me}{l}\right) = \frac{cl^{-1}}{CO3^{2-} + HCO3^{-} + SO4^{2-} + Cl^{-} + NO3^{-}}$$

Chloride concentration (me L^{-1})	Water quality
4	Excellent Water
4 – 7	Moderately good water
7 – 12	Slightly usable
12 – 20	Not suitable for irrigation purposes
> 20	

7. Soluble Sodium Percentage (SSP)

Among the soluble constituents of irrigation water, sodium is often regarded as the most hazardous. Elevated levels of sodium ions can render the water saline or alkaline, depending on whether they are associated with chloride or sulfate ions (saline) or with carbonate or bicarbonate ions (alkaline). Historically, the quality of irrigation water was assessed primarily based on sodium content, using a measure known as the Soluble Sodium Percentage (SSP). This metric was calculated to determine the proportion of sodium relative to other soluble ions, providing insight into the potential risks of sodium-related issues such as soil salinization or alkalinity.

8. Magnesium Hazard

Magnesium content in irrigation water is a critical factor in assessing water quality due to its impact on soil properties. The ratio of magnesium (Mg) to total divalent cations (calcium and magnesium) in irrigation water is an important qualitative criterion. High levels of magnesium can negatively affect soil structure by increasing soil dispersion, which in turn impacts water infiltration and soil aeration.

A detrimental effect on soil typically occurs when the calcium (Ca) to magnesium (Mg) ratio falls below 50. When the magnesium to calcium (Mg) ratio exceeds one, the irrigation water is considered to pose a "magnesium hazard." This imbalance can lead to adverse changes in soil physical properties, affecting overall soil health and crop productivity.

9. Nitrate Concentration (me L^{-1}):

Groundwater often contains elevated levels of nitrate. When irrigation water with high nitrate concentrations is applied continuously to soils, it can adversely affect various physical properties of the soil. This can lead to reduced plant growth and overall poor crop performance.

Water class	Nitrate Value (me L^{-1})	Remarks
Low Nitrate	5.0	Good water No problem
Medium Nitrate	5.0 – 30.0	Moderately good water
High Nitrate	> 30.0	Unsuitable for irrigation purposes

10. Lithium

Lithium is a trace element that can be present in saline groundwater and irrigated soils. In irrigation water, even small concentrations of lithium (0.05-0.1 ppm) cantmon crops, including rice, wheat, and barley, are generally not affected by these lithium levels during germination.

11. Fluoride

Fluoride is only sparingly soluble and typically occurs in small amounts in natural waters. Its concentration can range from traces up to more than 10 mg/L. In surface waters, fluoride concentrations rarely exceed 0.3 mg/L unless pollution is present. Studies have shown that irrigation with saline water containing fluoride concentrations up to 25 mg/L does not significantly impact wheat yield. Consequently, in India, fluoride levels are currently not considered a major concern, as average concentrations have not been observed to exceed 10 mg/L.

The World Food and Agriculture Organization (FAO) has established guidelines for irrigation water quality to ensure the effective and sustainable use of water resources in agriculture. These guidelines are designed to help manage water quality and minimize negative impacts on crops, soils, and the environment. Key aspects of the FAO's irrigation water quality guidelines include:

Water Constituent	Intensity of Problem*		
	No problem	Moderate	Severe
Salinity (dSm^{-1})	< 0.75	0.75 – 3.0	> 3.0
Permeability (rate of infiltration affected) Salinity (dSm^{-1})	> 0.5	0.5 – 0.2	< 0.2
Adjusted SAR : Soils are Dominantly montmorillonite Dominantly illite – vermiculite Dominantly kaolinite – sesquioxide	< 6.0 < 8.0 < 16.0	6.0 – 9.0 8.0 – 16.0 16.0 – 24	> 9.0 > 16.0 > 24.0
Specific ion toxicity Sodium (as adjusted SAR) Chloride (meq L^{-1}) Boron (meq L^{-1})	< 3.0 < 4.0 < 0.75	3.0 – 9.0 4.0 – 10.0 0.75 – 2.0	> 9.0 > 10.0 > 2.0
Miscellaneous $NO3^{-}$–N or NH^{+}–N (meq L^{-1}) 4 $HCO3^{-}$ (meq L^{-1}) as damaged by overhead sprinkler pH	< 5.0 < 1.5 6.5 -8.4	5.0 – 30.0 1.5 – 8.5 -	> 30.0 > 8.5 0.0 – 5.0, 9.5+

Note: Assuming that the soils range from sandy loam to clay loam, possess good drainage, and are located in arid to semi-arid climates, with irrigation methods being either sprinkler or surface, and root depths are typical for deep soils, the guidelines provided are approximate and should be adjusted based on specific conditions.

Crop Groups Based on Response to Soil Salinity

Sensitive Crops		Resistant Crops	
Highly sensitive	**Medium sensitive**	**Medium tolerant**	**Highly tolerant**
Lentil	Radish	Spinach	Barley
Mash	Cow pea	Sugarcane	Cotton
Chickpea	Broad bean	Indian mustard	Sugar beet
Beans	Vetch	Rice (transplanted)	Turnip
Peas	Cabbage	Wheat	Tobacco
Carrot	Cauliflower	Pearl millet	Safflower
Onion	Cucumber	Oats	Rapeseed
Lemon	Gourds	Alfalfa	Karnal grass
Orange	Tomato	Blue panic grass	Date palm
Grape	Sweet potato	Para grass	Ber
Peach	Sorghum	Rhodes grass	*Mesquite*
Plum	Minor millets	Sudan grass	*Casuarina*
Pear	Maize	Guava	*Tamarix*
Apple	Clover, *berseem*	Pomegranate	*Salvadora*

Upper Permissible Concentrations of Trace Elements in Irrigation Water

Element	**Recommended Maximum Concentration (mg/L)**	**Remarks**
Aluminium (Al)	5.0	Can cause non-productivity in acid soils (pH < 5.5), but more alkaline soils at pH > 7.0 will precipitate the ion and eliminate any toxicity.
Arsenic (As)	0.10	Toxicity to plants varies widely, ranging from 12 mg/l for Sudan grass to < 0.05 mg/l for rice.
Beryllium (Be)	0.10	Toxicity to plants varies widely, ranging from 5 mg/l for kale to 0.5 mg/l for bush beans.
Copper (Cu)	0.20	Toxic to a number of plants at 0.1 to 1.0 mg/l in nutrient solutions.
Fluoride (F)	1.0[3]	Inactivated by neutral and alkaline soils.
Iron (Fe)	5.0	Can be contributed to soil acidification and loss of availability of phosphorus and molybdenum.
Lithium (Li)	2.5	Tolerated by most crops up to 5 mg/l; mobile in soil. Acts similarly to boron.
Manganese (Mn)	0.20	Toxic to a number of crops but usually only in acid soils.
Molybdenum (Mo)	0.01	Can be toxic to livestock if forage is grown in soils with high concentrations of available Mo.
Nickel (Ni)	0.20	Toxic to a number of plants at 0.5 mg/l to 1.0 mg/l; reduced toxicity at neutral or alkaline pH.
Lead (Pb)	5.0	Can inhibit plant concentrations.
Zinc (Zn)	2.0	Toxic to many plants at widely varying concentrations; reduced toxicity at pH > 6.0 and in fine textured or organic soils.
Cadmium (Cd)	0.01	Toxic to beans, beets and turnips at concentrations as low as 0.1 mg/l in nutrient solutions. Conservative limits recommended it may be harmful to humans.

Conclusion

The quality of irrigation water is a critical factor that directly influences soil health, crop productivity, and long-term agricultural sustainability. Poor-quality water, containing excessive salts, heavy metals, or toxic elements, can lead to soil degradation, reduced crop yields, and ecological imbalances. Key parameters such as electrical conductivity (EC), sodium adsorption ratio (SAR), residual sodium carbonate (RSC), and the presence of harmful ions (e.g., chloride, boron) must be carefully monitored to assess water suitability. Effective management strategies, including leaching, blending with good-quality water, and adopting salt-tolerant crops, can mitigate adverse effects. Additionally, modern techniques like desalination and phytoremediation offer promising solutions for utilizing marginal-quality water. Policymakers and farmers must prioritize regular water testing, adopt sustainable irrigation practices, and implement proper drainage systems to prevent secondary salinization. Ultimately, ensuring high-quality irrigation water is essential for maintaining food security, preserving soil fertility, and promoting environmental conservation. By integrating scientific knowledge, technological advancements, and responsible water management, agriculture can thrive even in challenging conditions, securing a sustainable future for generations to come.

References

Adamu, G. K. (2013). Quality of irrigation water and soil characteristics of Watari irrigation project. American journal of engineering research, 2(3), 59-68.

Al-Omran, A. M., Al-Harbi, A. R., Wahb-Allah, M. A., Nadeem, M., & Al-Eter, A. (2010). Impact of irrigation water quality, irrigation systems, irrigation rates and soil amendments on tomato production in sandy calcareous soil. Turkish Journal of Agriculture and Forestry, 34(1), 59-73.

Anyango, G. W., Bhowmick, G. D., & Bhattacharya, N. S. (2024). A critical review of irrigation water quality index and water quality management practices in micro-irrigation for efficient policy making. Desalination and Water Treatment, 100304.

Arshad, M., & Shakoor, A. (2017). Irrigation water quality. Water Int, 12(1-2), 145-160.

Bauder, T. A., Waskom, R. M., Davis, J. G., & Sutherland, P. L. (2011). Irrigation water quality criteria (pp. 10-13). Fort Collins: Colorado State University Extension.

García-Garizábal, I., & Causapé, J. (2010). Influence of irrigation water management on the quantity and quality of irrigation return flows. Journal of Hydrology, 385(1-4), 36-43.

Hussain, G., Alquwaizany, A., & Al-Zarah, A. (2010). Guidelines for irrigation water quality and water management in the Kingdom of Saudi Arabia: an overview.

Irfan, M., Arshad, M., Shakoor, A., & Anjum, L. (2014). Impact of irrigation management practices and water quality on maize production and water use efficiency. JAPS: Journal of Animal & Plant Sciences, 24(5).

Jang, C. S., Chen, S. K., & Kuo, Y. M. (2012). Establishing an irrigation management plan of sustainable groundwater based on spatial variability of water quality and quantity. Journal of Hydrology, 414, 201-210.

Lothrop, N., Bright, K. R., Sexton, J., Pearce-Walker, J., Reynolds, K. A., & Verhougstraete, M. P. (2018). Optimal strategies for monitoring irrigation water quality. Agricultural Water Management, 199, 86-92.

Malakar, A., Snow, D. D., & Ray, C. (2019). Irrigation water quality—A contemporary perspective. Water, 11(7), 1482.

Orlob, G. T., & Woods, P. C. (1967). Water-quality management in irrigation systems. Journal of the Irrigation and Drainage Division, 93(2), 49-66.

Oster, J. D. (1994). Irrigation with poor quality water. Agricultural water management, 25(3), 271-297.

Ouda, S. A., Noreldin, T., Mounzer, O. H., & Abdelhamid, M. T. (2015). CropSyst model for wheat irrigation water management with fresh and poor quality water. Journal of Water and Land Development, (27), 41-50.

Qadir, M., Sposito, G., Smith, C. J., & Oster, J. D. (2021). Reassessing irrigation water quality guidelines for sodicity hazard. Agricultural Water Management, 255, 107054.

Torrion, J. A., & Stougaard, R. N. (2017). Impacts and limits of irrigation water management on wheat yield and quality. Crop Science, 57(6), 3239-3251.

Wichelns, D. (2002). An economic perspective on the potential gains from improvements in irrigation water management. Agricultural Water Management, 52(3), 233-248.

Wilcox, L. V., & Durum, W. H. (1967). Quality of irrigation water. Irrigation of agricultural lands, 11, 104-122.

Zaman, M., Shahid, S. A., Heng, L., Zaman, M., Shahid, S. A., & Heng, L. (2018). Irrigation water quality. Guideline for salinity assessment, mitigation and adaptation using nuclear and related techniques, 113-131.

11

Soil Pollution and Its Control

Introduction

Soil is considered as a thin layered living entity over the earth's surface, poses a profound implication on the solidity of ecosystems. It act as a transient nutrient sink for the plants and plays a crucial role in the crux of biogeochemical cycling of Earth's elements which is being carried out by distinct microbial assemblages present in the soil profile. Soil is the most diverse niche for the microbial communities in which large proportion of taxa exists at a very low correlative abundance. Despite the extensive importance of this entity it has been snubbed lately with the various undesirable activities done by human kind. Therefore, the scientific understanding of soil origin is critical. Scientifically speaking, formation of one inch of soil almost took 15 years if the parent materials are soft whereas hard parent material might takes hundreds of year for its development. Soil horizon involves the arrangement of mature soil in a series of zones. Each horizon encompasses the distinct texture and composition that varies with different kinds of soils. Soil profile is the cross sectional view of the horizons in a soil. The O horizon is the topmost layer of soil, mainly consists of partially decomposed leaves, animal wastes, fungi and other organic materials. **A** horizon is looser than other deeper layers of soil, mainly dark in colour and considered as the uppermost layer of the soil consisting of partially decomposed organic matter known as humus. Generally, in the upper layers of soil the roots of most plants exist. As long as these layers are supported by vegetation, it acts as a storehouse of water which gets released in a trickle throughout the year. Plant roots grow in the soil and absorb nutrients and water which are required during the process of photosynthesis that helps in the growth and development of plants. Soil contains essential nutrients which are being adsorbs by the plants and is beneficial to all living organisms such as herbivores and omnivores depends on plants for their food, then these being eaten up by carnivores and omnivores in the food web, so nutrients that are absorbed from the soil is reached to all strata of living entity. Soil is a niche for different organisms as its top layers harbors a large population of bacteria, fungi, earthworms and many other small insects that forms an important component of complex food webs in the soil that helps in recycling of soil nutrients.

Nowadays, this entity *i.e.* soil is being polluted due to the continuous building-up of recalcitrant toxic compounds, chemicals, salts, radionuclide etc. which imposes adverse effects on the growth of plant and also affects the animal health. The sources of this pollutant are the mining industries, various manufacturing units and their effluents contain heavy metals which are considered as the most serious pollutants causing land pollution. However, the risk assessment focus is exclusively on heavy metal pollution for which spatial distribution of pollutants, its mobility in soils provides important information. Organic pollutant is responsible for the organic pollution, which is being generated by the petrochemical industries, despite its strong presence in the industrial and mining gathering areas and looking to its potential harm, most of the time it has not been addressed properly. Therefore, a comprehensive approach considering the soil risk which is being caused by the complex pollutants, is needed to get acquaint with the overall status of the soil environment especially, the soil near industrial and mining gathering areas. Improper anthropogenic activities or the reckless act of mankind like littering of plastic, tin, glass, dumping of toxic waste, deforestation *etc* abuses the land in turn, harms the soil and due to this all the living entity requiring fertile soil to survive gets affected. Also, it is a major hindrance to the societal development which needs to be checked immediately.

Source of Soil Pollution

Soil pollution is caused due to:

1. Industrial wastes
2. Urban wastes
3. Agricultural wastes
4. Radioactive wastes

1. Industrial wastes

Disposal of waste from industries is one of the main reasons of soil pollution. Pollutant of various industries are mainly get released from multiple points of operation like paper and pulp industry, manufactured fertilizer industry, oil refineries, sugar mills, textile industry, steel and Iron industries, coal and mineral mining industries, petroleum industries etc. Improper handling and disposal of these pollutants contaminates the soil which ultimately destroy the soil profile in the long run.

With urbanization, the standard of living of people is enhanced and there is consistent clamor for the need of various good and services. To satiate the demands of people the activities relating to manufacturing have shot up and astronomically, the number of manufacturing units and industries is increased and are many more are still lined up.

As a result, various industrial activities lead to generation of industrial wastes which ultimately pollute the land.

2. Urban wastes

Urban wastes encompass both *i.e.* waste which are commercially and domestically generated, comprising of dried sewage and sludge. The urban refusal consisting of garbage and waste materials like non-biodegradable plastics, glass wares, metal and tin cans, paper bits, rubber materials, fuel residues, deserted vehicles and other rejected manufactured products which are xenobiotics and recalcitrant in nature. The exercise of dumping is immanently prevalent in the countries which are economically backward or under developed.

The disposal and treatment of urban domestic wastes is different from other industrial wastes however, still poses more or less same risk this is because of non-biodegradability of urban wastes.

3. Agricultural wastes

An agricultural waste mainly involves the fabricated fertilizers and pesticides that pollute the soil. The use of fertilizer is done to increase the productivity of crop; however, its application above the recommended dose leads to leaching of nutrients present in it from the fields to the nearby water bodies and causes eutrophication in the aquatic system. The pesticides are applied to control the attack of pest on crop however; it also contains chemicals that are carcinogenic and toxic, and imposes a negative impact on human and animal's health.

4. Radioactive waste

The nuclear power plant generates a waste called a radioactive waste. Owing to the prudent importance of nuclear plants, many countries prefer to discharge their radioactive waste directly into the ocean or sea without being treated. If these nuclear wastes in manifolds get accumulated at that specific spot of a water body then such water body no longer supports the different forms of life. If at this point also, immediate measures is not being taken then it gets reached to the sea shore. This kind of disposal but not proper removal at the same time leads to its accumulation in the soil atmosphere. From the past few years people sees radioactive waste with great concern and for good reason because at any concentrations it is harmful for the living organisms whether it present in low or high quantity. Examples of such kind of wastes are radio nuclides of Strontium, Radium, Uranium, Thorium, Uranium *etc.* constitutes the radioactive wastes.

Effects of Soil Pollution

i. Effects on human health

Soil pollution imposes a serious effect on the wellbeing of whole society whether it is of urban or rural. The soil is contaminated mainly with the recurrent heavy metals like cobalt, nickel, copper, zinc, arsenic, chromium, silver, cadmium, antimony, selenium mercury, thallium, fluoride and lead *etc.* which are carcinogenic in nature and its chronic exposure will give rise to various congenital diseases or birth defects like cleft palate in the new borne, aplasia of lungs *etc.* and many other chronic health problems like bone deformities, cancer *etc.* Ammonia and nitrates considered as the main ingredients of most of the fertilizers, and are being identified as dangerous compounds deteriorating the health and quality of soil due to its direct soil application. Higher concentration of these heavy metals in the soil causes serious ailments such as kidney damage, liver toxicity, low IQ in children that is being associated with the higher concentration of mercury, cyclodienes, PCBs, lead, benzenes *etc.* in the soil. Other compounds like carbamates and organophosphates affect the nervous system and cause many other neurological disorders in humans.

Major consequences of soil pollution on human health are summed up below

- Genetic makeup of the body gets disturbed due to the chronic exposure to pollutants
- Long term exposure to heavy metals, petroleum industry solvents and agrochemicals products will lead to the cancer. As leukemia a type of blood cancer is associated with the long term exposure to benzene compounds.
- Higher concentration of mercury in the atmosphere leads to Kidney damage.
- Liver toxicity is associated with the cyclodienes compounds.
- Neuromuscular breakdown occurs due to organophosphates compounds.

Effects on Animals

- Bioaccumulation of soil contaminant in the body of animals affects their metabolic versatility in the long run.
- As the lower living entities may feed on toxic pollutant present in the soil which may then be rolled up to the food chain to larger forms of life; that not only can demolish some of the layers of the primary food chain, but also leads to the extinction of larger forms of life *i.e.* of animals as well.

Effect on Agriculture

- Soil pollution leads to slumping of soil fertility which is the most evident and crucial component of soil. Chemicals and heavy metals depreciate the fertility by denaturing the active enzymes that revitalizes the soil.
- It leads to alteration in structure of soil which is due to the multiple factors like acidification and decreased fertility of soil leads to the reduction in microbial population. As breakdown of organic matter is favored by the microorganism present in the soil and this breakdown enhances the penetration and retention capacity of soil thereby promoting the soil structure.
- Due to improper irrigation and various other agricultural practices, deposition of nitrates and phosphate occur in the soil which ultimately increases the salt concentration. In the presence of excess salts, roots of plant unable to absorb the essential nutrients and moisture from the soil.
- Due to salinization, acidification etc. the rhizobacteria growth gets affected thereby, restricting the plant growth promoting activities in the soil.

Effects on ecosystem

- The whole soil chemistry gets changed with the soil pollution, affecting the integrity of ecosystem.
- With this, there is a quick change in chemical nature of soil and the whole vegetation gets affected because the plants won't be able to acclimatize in such a short span of time. Also, the microbial load of soil such as the population of fungi and bacteria get decreased.
- Soil pollution is also associated with the food chain, operating at primary level that could lead to complete exclusion of lower living forms of the whole biosphere. Therefore, the ecological balance of the whole system gets disturbed.

Control measures associated with soil pollution

Some of the measures suggested controlling the soil pollution

1. **Minimizing the use of fertilizer and pesticide:** Crop grown under the conventional system patently, ensures the food security but also is the major source of pollution, deteriorating the quality of environment. The commonly used pesticides are DDT, BHC, aldrin, organophosphates etc. poses residual effects in soil. Soil particles adsorb the remnants of these pesticides thereby, contaminating the crops cultivated on that soil.

Adoption of the organic method for the cultivation of major crops is one such alternative which employs the use of bio-fertilizers and biomanure instead of chemical fertilizer and pesticide. Also, Biological control methods for pest control reduces the use of pesticides thereby lessen the risk of soil pollution.

2. **Reuse of materials:** Some of the daily use materials such as glass wares, polythene bags, paper, cloth etc. could be reused at the domestic levels rather than being discarded in the environment, thereby, minimizing the soil pollution.
3. **Recycling and recovery of materials:** This is considered one of the reasonable means for reducing the soil pollution. Some of the routinely used stuffs like paper, plastics, glass and tin cans could be recycled. This helps in minimization of the volume or load of the discarded materials in the environment thereby conserving the most important entity of earth *i.e.* soil.
4. **Reforesting:** Through attempting the practice of reforestation, the soil loss through erosion can be checked. Practicing crop rotation or mixed cropping enhances the soil fertility thereby improving the quality of soil.
5. **Solid waste treatment:** The burgeoning urban population of underdeveloped countries made the solid waste management an important issue to be considered. It has been seen that the city authorities spends a significant amount of total expenditures on collecting the solid waste. Solid waste management is done through land filling, pyrolysis, composting and incineration which are considered expensive in terms of their operation. Therefore, Optimizing the routing system for collection and transporting the solid waste constitutes an important element for an effective solid waste treatment. A model has been devised recently which is known as GIS optimal routing model which will evaluate the minimum cost for the distance efficient collection paths for the transportation of the solid wastes to the landfill.

Conclusion

Soil forms the base for the very human existence and development since all living forms on this earth directly or indirectly dependent on it for their nourishment. If the, soil not being used strategically or its pollution continues unabated then the economic prosperity in turn, societal development of a country will be doomed to face the sulkiness. Much research and technological advances are needed to rein the soil pollution. Farmers are adopting more conventional approach *i.e.* organic way of farming to invert the disfigurement caused to the soil in the context of structure and its composition. Hardier native plants are now being cultivated in exchange of the hybrid ones. As the awareness of healthy soil environment grows the people must pitch in by considering the motto of three R's *i.e.* recycle, reduce and reuse. Therefore, consumerist which prevails in us must reduce and the conservationist must improve to shield and protect the soil.

References

Adie, G.U. and Osibanjo, O., 2009. Assessment of soil-pollution by slag from an automobile battery manufacturing plant in Nigeria. African Journal of Environmental Science and Technology, 3(9).

Beesley, L., Moreno-Jiménez, E. and Gomez-Eyles, J.L., 2010. Effects of biochar and greenwaste compost amendments on mobility, bioavailability and toxicity of inorganic and organic contaminants in a multi-element polluted soil. Environmental pollution, 158(6), pp.2282-2287.

Blume, H.P., Brümmer, G.W., Fleige, H., Horn, R., Kandeler, E., Kögel-Knabner, I., Kretzschmar, R., Stahr, K. and Wilke, B.M., 2016. Soil Development and Soil Classification. In Scheffer/SchachtschabelSoil Science (pp. 285-389). Springer Berlin Heidelberg.

Cheng, M., Zeng, G., Huang, D., Lai, C., Xu, P., Zhang, C. and Liu, Y., 2016. Hydroxyl radicals based advanced oxidation processes (AOPs) for remediation of soils contaminated with organic compounds: a review. Chemical Engineering Journal, 284, pp.582-598.

Dasgupta, P.K., Dyke, J.V., Kirk, A.B. and Jackson, W.A., 2006. Perchlorate in the United States. Analysis of relative source contributions to the food chain. Environmental science & technology, 40(21), pp.6608-6614.

Datta, S., Singh, J., Singh, S. and Singh, J., 2016. Earthworms, pesticides and sustainable agriculture: a review. Environmental Science and Pollution Research, 23(9), pp.8227-8243.

Doran, J.W. and Zeiss, M.R., 2000. Soil health and sustainability: managing the biotic component of soil quality. Applied soil ecology, 15(1), pp.3-11.

Duruibe, J.O., Ogwuegbu, M.O.C. and Egwurugwu, J.N., 2007. Heavy metal pollution and human biotoxic effects. International Journal of Physical Sciences, 2(5), pp.112-118.

Fetter, C.W., Boving, T. and Kreamer, D., 2017. Contaminant hydrogeology. Waveland Press.

Frangi, J.P. and Richard, D., 1997. Heavy metal soil pollution cartography in northern France. Science of the Total Environment, 205(1), pp.71-79.

Ghose, M.K., Dikshit, A.K. and Sharma, S.K., 2006. A GIS based transportation model for solid waste disposal–A case study on Asansol municipality. Waste management, 26(11), pp.1287-1293.

Gong, Y., Tang, J. and Zhao, D., 2016. Application of iron sulfide particles for groundwater and soil remediation: a review. Water research, 89, pp.309-320.

Guan, Y., Shao, C., Gu, Q., Ju, M. and Zhang, Q., 2015. Method for assessing the integrated risk of soil pollution in industrial and mining gathering areas. International journal of environmental research and public health, 12(11), pp.14589-14609.

Harrison, R.M. ed., 2001. Pollution: causes, effects and control. Royal Society of Chemistry.

Himeno, S. and Imura, N., 2002. Selenium in nutrition and toxicology. Heavy Metals in the Environment. Marcel Dekker Inc. New York, NY, pp.587-629.

Ho, Y.S. and McKay, G., 1998. A comparison of chemisorption kinetic models applied to pollutant removal on various sorbents. Process safety and environmental protection, 76(4), pp.332-340.

Huguenot, D., Mousset, E., Van Hullebusch, E.D. and Oturan, M.A., 2015. Combination of surfactant enhanced soil washing and electro-Fenton process for the treatment of soils contaminated by petroleum hydrocarbons. Journal of environmental management, 153, pp.40-47.

Kadi, M.W., 2009. "Soil Pollution Hazardous to Environment": A case study on the chemical composition and correlation to automobile traffic of the roadside soil of Jeddah city, Saudi Arabia. Journal of hazardous materials, 168(2), pp.1280-1283.

Kanjana, D., 2017. Advancement of Nanotechnology Applications on Plant Nutrients Management and Soil Improvement. In Nanotechnology (pp. 209-234). Springer, Singapore.

Kozlov, M.V., Zvereva, E.L. and Zverev, V.E., 2009. Soil Quality. In Impacts of Point Polluters on Terrestrial Biota (pp. 107-131). Springer Netherlands.

Li, Z., Ma, Z., van der Kuijp, T.J., Yuan, Z. and Huang, L., 2014. A review of soil heavy metal pollution from mines in China: pollution and health risk assessment. Science of the Total Environment, 468, pp.843-853.

Mao, X., Jiang, R., Xiao, W. and Yu, J., 2015. Use of surfactants for the remediation of contaminated soils: a review. Journal of hazardous materials, 285, pp.419-435.

Mirsal, I., 2008. Soil pollution: origin, monitoring & remediation. Springer Science & Business Media.

Misra, S.G. and Mani, D., 1991. Soil pollution. Ashish Publishing House.

Monte, M.C., Fuente, E., Blanco, A. and Negro, C., 2009. Waste management from pulp and paper production in the European Union. Waste management, 29(1), pp.293-308.

Mousset, E., Oturan, N., Van Hullebusch, E.D., Guibaud, G., Esposito, G. and Oturan, M.A., 2014. Influence of solubilizing agents (cyclodextrin or surfactant) on phenanthrene degradation by electro-Fenton process–Study of soil washing recycling possibilities and environmental impact. water research, 48, pp.306-316.

Nahmani, J., Lavelle, P., Lapied, E. and van Oort, F., 2003. Effects of heavy metal soil pollution on earthworm communities in the north of France: The 7th international symposium on earthworm ecology· Cardiff· Wales· 2002. Pedobiologia, 47(5-6), pp.663-669.

Pan, J., Plant, J.A., Voulvoulis, N., Oates, C.J. and Ihlenfeld, C., 2010. Cadmium levels in Europe: implications for human health. Environmental geochemistry and health, 32(1), pp.1-12.

Peluffo, M., Pardo, F., Santos, A. and Romero, A., 2016. Use of different kinds of persulfate activation with iron for the remediation of a PAH-contaminated soil. Science of the Total Environment, 563, pp.649-656.

Rowell, D.L., 2014. Soil science: Methods & applications. Routledge.

Salomons, W., Forstner, U. and Mader, P. eds., 2012. Heavy metals: problems and solutions. Springer Science & Business Media.

Schimel, D.S., 1995. Terrestrial ecosystems and the carbon cycle. Global change biology, 1(1), pp.77-91.

Smith, O.H., Petersen, G.W. and Needelman, B.A., 1999. Environmental indicators of agroecosystems. Advances in agronomy, 69, pp.75-97.

Stewart, B.W.K.P. and Wild, C.P., 2017. World cancer report 2014. Health.

Trasar-Cepeda, C., Leiros, M.C., Seoane, S. and Gil-Sotres, F., 2000. Limitations of soil enzymes as indicators of soil pollution. Soil Biology and Biochemistry, 32(13), pp.1867-1875.

Trellu, C., Mousset, E., Pechaud, Y., Huguenot, D., Van Hullebusch, E.D., Esposito, G. and Oturan, M.A., 2016. Removal of hydrophobic organic pollutants from soil washing/ flushing solutions: a critical review. Journal of hazardous materials, 306, pp.149-174.

Van Elsas, J.D., Trevors, J.T., Jansson, J.K. and Nannipieri, P. eds., 2006. Modern soil microbiology. CRC press.

Van Straalen, N.M. and Roelofs, D., 2008. Genomics technology for assessing soil pollution. Journal of Biology, 7(6), p.19.

Wang, Y., Shi, J., Wang, H., Lin, Q., Chen, X. and Chen, Y., 2007. The influence of soil heavy metals pollution on soil microbial biomass, enzyme activity, and community composition near a copper smelter. Ecotoxicology and environmental safety, 67(1), pp.75-81.

Waseem, A., Arshad, J., Iqbal, F., Sajjad, A., Mehmood, Z. and Murtaza, G., 2014. Pollution status of Pakistan: a retrospective review on heavy metal contamination of water, soil, and vegetables. BioMed research international, 2014.

Wilcke, W., 2000. Synopsis polycyclic aromatic hydrocarbons (PAHs) in soil—a review. Journal of plant nutrition and soil science, 163(3), pp.229-248.

Wu, H., Lai, C., Zeng, G., Liang, J., Chen, J., Xu, J., Dai, J., Li, X., Liu, J., Chen, M. and Lu, L., 2017. The interactions of composting and biochar and their implications for soil amendment and pollution remediation: a review. Critical reviews in biotechnology, 37(6), pp.754-764.

Xu, Y. and Zhou, N.Y., 2017. Microbial remediation of aromatics-contaminated soil. Frontiers of Environmental Science & Engineering, 11(2), p.1.

Yang, H., Huang, X., Thompson, J.R. and Flower, R.J., 2014. Soil pollution: Urban brownfields. Science, 344(6185), pp.691-692.

Zhai, M., Kampunzu, H.A.B., Modisi, M.P. and Totolo, O., 2003. Distribution of heavy metals in Gaborone urban soils (Botswana) and its relationship to soil pollution and bedrock composition. Environmental Geology, 45(2), pp.171-180.

Zhang, J. and Liu, C.L., 2002. Riverine composition and estuarine geochemistry of particulate metals in China—weathering features, anthropogenic impact and chemical fluxes. Estuarine, coastal and shelf science, 54(6), pp.1051-1070.

12

Water Pollution and Its Management

Introduction

Water is chemical substance which important for all living organism to survive on this planet. Water is necessary for all cell of the living organism's body to performance any normal work. Water covers 71% of the Earth's surface, most like in oceans and other more water bodies, with 1.6% of water below ground in aquifers and 0.001% in the air as vapors, clouds and precipitation. Some observation have estimated that by 2025 more than half of the world population will be facing water-based vulnerability, a situation which has been called a 'water crisis' by the United Nations. A recent report suggests that by 2030, in some developing regions of the world, water desire will be exceed supply by 50%. The total volume of water 1,385 million Km^3 on the planet earth, 96.5% is salt water (oceans and seas), the fresh water is mostly ice (24 million Km^3). The fresh water available as annual stream flow is 46,768 Km^3, that is 0.00034% of the total global water. If there is no intake of water into the body, death can ensue in 7-10 days. Water is also essential to man for maintaining personal hygienic and freedom from disease. Today there are many cities worldwide facing actual shortage of water and nearly 40 percent of the world's food supply is grown under irrigation and a wide variety of industrial processes depends on water. Water is the most vital element among the natural resources, and is critical for the survival of all living any organisms including human, food production, and economic development. The environment, economic growth and developments are highly affected by water is regional and seasonal availability and the quality of surface and groundwater. The quality of water is affected by any human activities and is declining due to the rise of urbanization, population growth, industrial production, climate change and other factors. The resulting water pollution is a serious surrounding to the well-being of both the Earth and its population. The pressures is increasing population, growth of industries, urbanization, energy intensive life style, loss of forest cover, lack of environmental awareness, lack of implements of environmental rule, regulations and environment improvement plans, untreated effluent discharge from industries and municipalities, use of non-biodegradable pesticides/fungicides/ herbicides/insecticides, uses of chemical fertilizers few amount of organic manures, etc are causing water pollution. The pollutants from industries discharge and sewage besides finding their way to surface water reservoirs and rivers are also percolating into ground to pollute ground water sources. The polluted water may have undesirable colour, odor, taste, turbidity, organic matter contents, harmful chemical contents, toxic and heavy metals, pesticides, oily matters, industrial waste products, radioactivity, high Total Dissolved Solids (TDS), acids, alkaline, domestic sewage content, virus, bacteria, protozoa, rotifers, worms, etc. The organic content may be biodegradable or non-biodegradable. Pollution of surface waters (rivers, lakes and ponds), ground waters are all harmful for human and animal health. Pollution of the drinking water and that of food chain is by far the most worry-some aspect. In order to avoid ill effects of water pollution on the human, animal health and agriculture, standards/rules/guidelines have been devised for discharge of effluents from industries and municipalities, quality of drinking water, irrigation water, criteria for aquatic life in fresh water by various authorities including central pollution control board (India), World Health Organization (WHO), World Bank, Indian Standard Institution, Indian Council of Medical Research, etc. The implementation of these rules, standards and guidelines, etc is however, wanting. Alarming level of lead (Pb) and arsenic (As) has been found in the sediments as well as waters of Damodar, Safi, Ganga, Adjai rivers in Jharkhand, and West Bengal. High level of contamination by heavy metals, chemicals, organic matter, nitrates, coliforms, human and animal excreta, pesticides, etc is found in various rivers in India including Ganga, Yamuna, Gomti, Ramganga, Hindon, Chambal, Godavari, Krishna, Sabarmati, Subernrekha, Cauvery, etc specially near big cities and industries. Similar is the condition of lakes and ponds near cities.

"Pollution is defined as a natural change in the quality of water which render it unusable as regards food, human and animal health, industries, agriculture, fishing or leisure pursuits."

Utility of water

Water is desire for different other personal uses namely:

1. Public water supplies

Water is used for domestic purpose like as:

- Drinking purpose,
- Cooking and dishwashing purpose,
- General cleaning and laundry purpose,

- Personal washing, bathing and lavatory flushing purpose,
- Car washing purpose and
- Garden irrigation purpose.

2. Industrial water Supplies

Different types industrial are requires more quantities of water. They are given below:

- Manufacturing products.
- Steam raising
- Material processing
- Cooking purpose
- Solvent purpose
- Chemical Manufacturing purpose

3. Cooling water

Water cooling is commonly used for cooling and they are given below:

- Automobile internal combustion engines purpose
- Steam electric power plants purpose,
- Hydroelectric generators purpose,
- Petroleum refineries purpose,
- Chemical plants purpose.

4. Water for Agricultural Industry

The agricultural is used of water different purpose and they are given below:

- Industry uses comparatively small quantities of water,
- Dairy processing purpose,
- Removal animal hygiene purpose,
- stock watering for animal drinking and irrigation purpose,
- Land irrigation,
- Horticultural industry uses water for land irrigation and
- Glasshouse watering and washing marketable vegetable

5. Water for amenity and recreational purposes

Different purpose uses of water:

- Sports such as swimming purpose,
- Fishing and boating purpose and
- Sailing and skiing of transport purpose.

Classification of water pollution

Water pollution is classified into two types:

1. Surface water pollution and
2. Ground water pollution.

1. Surface Water Pollution

When pollutants are entering in to stream, river or lake these give rises from surface water pollution. The surface water pollution has sources. These may be categorized into two parts:

i. Point and Non-point Sources and

ii. Natural and Anthropogenic Sources

i. Point and non-point sources

Point Sources are defined as the sources that throw pollutants or effluents directly into different water bodies of fresh water are called point sources such as domestic and industries wastes. The point sources of pollution are effectively checked. The other word, the non-point sources of water pollution are scattered large areas. This type of sources conferred pollutants indirectly care of environmental variation and causes for many of the contaminants in flow of lakes. For example, the effective water that flow from agriculture farms, construction place, enters streams and lakes.

ii. Natural and anthropogenic sources

Natural source is defined as the increase in the concentration of naturally occurring materials is also termed pollution. The sources of material increase are called natural sources. They are sources such as Siltation (soil, sand and mineral particles included with water) is a natural source. Natural source is a common natural phenomenon, which come out in most water parts. Reckless deforestation is texture when soil loose them flood with waters and bring silt from rivers, mountains into streams and lakes. Anthropogenic is defined as the when human activities that cause the pollution of water is called anthropogenic Ex. domestic (sewage and waste water), industrial and agricultural wastes and they goes into the rivers, lakes, streams and seas, are cause the pollution source of anthropogenic. Somewhat materials are leached out the soil by water and enter the some part of water bodies.

2. Ground water pollution

Ground water pollution is defined as the when the polluted water percolate into the ground or land is called ground water pollution. The pollution is mostly occurs in the villages and many town area, ground water is the single sources of drinking water. Therefore, pollution of groundwater is a serious matter. Groundwater is gets polluted in a different ways such as dumping of raw sewage material in the soil, pits and septic tanks; they are causes pollution of groundwater. The porous layers of soil adsorbed of more polluted water. The soluble pollutants water is efficient to mix with the groundwater. The excessive use of nitrogenous fertilizers and release of toxic wastes and even carcinogenic materials by industrial units many results in slow trickling down through the earth's surface and mixing with the groundwater. This problem is very serious especially in some areas where water table is high. In such a case it is difficult to estimate the source of water pollution.

Water pollutants

You have read the various sources from pollutants enter in the water bodies. These can be categorized in to different parts:

i. Sewage Pollutants,

ii. Industrial Pollutants,

iii. Agricultural Pollutants and

iv. Radioactive and Thermal Pollutants.

Sewage pollutants

- Sewage is defined as the wastewater that mostly contain faces, urine and laundry waste.
- The sewage contains are more pollutants such as soaps, detergents, waste food, garbage and human excreta they are largest causes of water pollution.
- Disease causing microorganisms are such as bacteria, fungi, protozoa and algae, enter in to the water system throw sewage making it infected.
- Water polluted by sewage is found few bacteria and viruses cannot grow but themselves are reproduces in the cells of host organisms.
- They are caused the disease such as polio, viral hepatitis and may be cancer.
- Sewage disposal is a serious problem in creating countries as mostly people in areas don't have enters to sanitary conditions and fresh water.
- Sewage is developed is carried from the home immediately and hygienically care of sewage pipes.
- Sewage is mainly biodegradable and most of it is broken down in the environment.

- Also sewage is often causes problems when people use the flush chemical and pharmaceutical materials down the toilet. And then cause the water pollution

ii. Industrial pollutants

Industries are using the freshwater to carry throw waste from the plant into rivers, oceans and lakes. These are responsible for discharging their untreated effluents into rivers like highly toxic heavy metals such as asbestos, lead, mercury, nitrate, phosphate, sulphur oil and petrochemical etc. River Ganges receives wastes from textile, sugar, paper pulp mills, tanneries, rubber, and pesticide industries.

Pollutants from industrial sources given below:

- Asbestos – Asbestos pollutant is a serious problem for health hazard. Asbestos fibers may be inhaled and cause disease like as asbestosis, mesothelioma, lung cancer, intestinal cancer and liver cancer.
- Lead – Lead is a metallic element and it can be cause health and environmental problems. Lead is a non-biodegradable material so is hard to clean up once the environment is muddy. Lead is obnoxiously for health of many animals and humans.
- Mercury – Mercury is a metallic element and can cause health and environmental problems. Mercury is mostly harmful to animal health. It may be cause disease through mercury poisoning.
- Nitrates – The highly use of nitrates fertilizers are washing in to the soil and water. This is manufacturing for the industries and sources the water pollution, which can be very problematic to rivers and marine environments.
- Phosphates – The increased use of phosphoric fertilizers and manufacturing for the different industries and causes the water pollution. This can cause eutrophication, which can be very problematic to marine environments.
- Sulphur – Sulphur is a non-metallic substance that is harmful for marine life.
- Oils – Oil is not dissolve in water, but it forms a thick layer on the water surface. This can stop marine plants receiving enough light for photosynthesis. It is also harmful for fish and marine birds.
- Petrochemicals – This is cause from gas or petrol and can be toxic to marine life.

iii. Agricultural waste

The most important use of water in agriculture is for irrigation, which is a key component to produce enough food. Agricultural pollution refers to biotic and abiotic byproducts of farming system and cause the water pollution.

Pesticides: More use of pesticide and herbicides are used in agriculture fields to control pests and diseases that why, pesticide persisting in nature to accumulate in soil to altered the microbial process, chemical increase in plants and also occur toxicity in plants and soil organisms and its can be cause the water pollution.

Fertilizers: Highly use of fertilizers in agriculture fields to increase the crop production. The fertilizers are accumulates in the soil and loss of throw runoff. The high application rate of fertilizers such as nitrogen-containing combine with soil and produce the water solubility to increases the surface run off. It leached in to the ground water. They are causing ground water pollution.

Cadmium: Cadmium is a phosphorus containing fertilizers varies considerably and produced the problematic such as mono-ammonium phosphate. Highly use of cadmium fertilizers can contaminate with soil and plants. It is leached with the water in the soil and source of water pollutants.

Fluoride: Fluoride is chemical substance and contain in phosphate rocks. When phosphate fertilizers are used in the soil and increase the fluoride concentrations. It has been found that food contamination from fertilizer is concern as plants accumulate fluoride from the soil, and greater possibility of fluoride toxicity to livestock's. Also are effects of fluoride on the soil microorganism.

Heavy metals: The major of heavy metals like as lead, cadmium, zinc, arsenic, mercury use in to agriculture systems are fertilizers, organic waste such as manures and industrial waste and forming system such as irrigation to accumulate of heavy metals in the soil. It is highly toxicity to wildlife, livestock and humans and source of water pollutants.

Leaching, runoff, and eutrophication: Highly chemical fertilizers are applied to agriculture land can provide the sufficient plants nutrients. When the excess of fertilizers and runoff surface on the soil and leached out and, cause ground water.

iv. Physical pollutants

Physical pollutants may be some different types. They are discussed below:

Radioactive wastes: They are radio nuclides found in the water radium and potassium-40. These are originated from the natural sources due to leaching from the minerals and accidental leakage of waste in to uranium and thorium mines, nuclear power plants, research laboratories and hospital any where use radioisotopes. The radioactive materials enter into water bodies and source of water pollutants.

Thermal sources: The different type of industries such as nuclear power plants and thermal plants are requires water for cooling purpose and hot water is discharged in to the rivers and lakes. This is resultants for thermal pollutants and cause to the imbalance in the ecosystem of water bodies.

Sediments: Soil particles flow to streams, lakes or oceans form the sediments. Soil erosion defined as the soil carried by flood water from crop land, is responsible for sedimentation. The sediments may cause the water body by introducing a large amount of nutrient matter and sources of water pollutants.

Petroleum Products: Petroleum products are mostly used for fuel, lubrication and plastics manufacturing. Crude oil and other related products generally get into water by accidental spillage from ships, tankers, pipelines etc. Besides these accidental spills, oil refineries, oil exploration sites and automobile service centers pollute different water bodies.

Micro-organisms causing water pollution

There are various type micro-organisms agents which are also cause water pollution if drinking water found contaminated with these agents. The micro-organisms is involved such as bacteria, viruses and protozoa which may cause diseases that vary in severity from mild gastroenteritis to severe and sometime fatal diarrhea, dysentery, hepatitis or typhoid fever. Example of micro-organisms which are source of water pollutants such as parasites (protozoans and helminthes), cholera (*Cholera vibrio*), shigellosis (dysentery caused by *Shigella* spp.), typhoid (*Samonella typhi*), paratyphoid (*Samonella paratyphi*), diarrhea (*Escherichia coli*), hepatitis (Hepatitis virus) and poliomyelitis (Polio virus).

Prevention and control of water pollution

There are different way to prevent and control of water pollution-

1. **Sewage treatment:** Sewage water have 99.9 % water and 0.1% solid matter. Domestic waste water is sewage water. Sewage water should not be directly disposed to water bodies. It should be treated. Sewage water treatment have three steps. Primary treatment is the first step. This is the physical process. In this treatment, water is treated by following way- Removal of grits and pebbles, screening, sedimentation, flocculation, coagulation, filtration. Second step is biological treatment of sewage water. Biological treatment of sewage water involve the aerobic and aerobic biological treatment.

2. **Solid waste management:** To control the water pollution we should be done proper management of solid waste. There are different methods of solid waste management like, sanitary land filling, composting, pyrolysis and incineration.
3. **Prevention of river water from contamination:** Cleaning of river water is not easy by natural process. So waste water from industry or sewage should not be added without treatment. Solid waste or liquid waste should not be thrown to river. Some external substances should be added to river to decontaminate of river water.
4. **Liquid waste treatment before discharge:** Effluent from factories or agriculture industry, refineries, should be treated prior to discharge. And treated effluent should try to recycle.
5. **Strictly follow the water pollution control law and order:** All people should follow the law and legislation with regards to pollution control.
6. **Drainage water treatment:** A large amount of water is drained every day in cities which contain various types of disease causing microorganisms and chemical substances. These are harmful to human and causes of water pollution if this meet with water bodies. So water flow that drainages from cities.
7. **Establishment of treatment plant:** Treatment plant should be established in metro cities and towns. These plants filter out undissolved material and other harmful chemical substances. So this treated water can be recycled to agriculture purpose.
8. **Village pond water should be maintained clean and safe:** Washing of clothes in pond and bathing of cattle in pond make it dirty. If these ponds are continually misuses, then it may lead of severe consequences.
9. **Don't throw insecticide box in sinks and toilets**
10. **Drinking water should be kept in more hygienic condition**
11. **Public awareness**

Effect of water pollution

The effects of water pollution can be under the following headings:

1. Physical effects.
2. Oxidation effects
3. Toxic chemical effects
4. Chemical nutrients effects
5. Pathogenic effects
6. Radionuclide effects

1. Physical effect

Physical effects are divided into the organic solid, undergo slow biodegradation and affect by a reduction of the soluble oxygen in the water. Defectives solids are of different particle sizes and density, and they are also settles out suspended according to based on the properties and the turbulence of the water. The settled particles can be few accumulate on vegetation part, and create a deposit on the river side. The affected of settlement particles layers are to decreases the solar energy absorption by plants and the rate of photosynthesis are produce low oxygen démodé on the river beds. Some suspended particles produce water turbidity and this reduces light penetration, decrease photosynthesis rate and control plant growth. Turbidity also decrease visibility in the water and limits the food catching capacity of more animals. Fish and some invertebrates have their respiratory efficiency decrease because the gills surfaces become clogged with suspended matter. All these physical effects caused a disturbed of the balanced ecosystem. Few animal species do not survive, others are decrease in numbers, and so food cycle and actual are affected. Waste oils, fats and grease may be produced rivers and estuaries from several water sources. Oils and allied petrol-chemicals form a thin film on the water surface which causes the exchangeable of oxygen with the atmosphere. This causes a decrease of the water oxygen saturation and deoxygenated pollution effects increase.

At sea, oil slicks are responded for the deaths of many birds eg. Scoter, puffins, razorbills and divided duck. When oil is deposited on rocks and sand, and this inhibited the beaches being used for generation and enjoyment by the public, and affects marine life. A badly oiled shore can be largely denuded of animal life, and sea weeds are also affected.

2. Oxidation effect

There are two types of oxidation:

i. Action of bacteria on organic pollutants
ii. Chemical oxidation of other pollutants present in industrial waste.

Both types of oxidation involve the use of soluble with the oxygen, and so product an increased BOD and an oxygen decrease in the water sources. Examples of bacteriological oxidation are: conversion of sulphide to sulphate in the sulphur cycle, and ammonia to nitrite and then nitrate in the nitrogen cycle.

3. Chemical toxic effect

Some organic and inorganic chemical materials are toxic to flora and fauna. Chemical toxins can be long ranged considered under the following four headings and they are given below:

i. Metal and salt toxins
ii. Pesticides toxin
iii. Acid and alkali toxins
iv. Other organic compounds toxins

i. **Toxic metals:** These are mostly heavy metals like as Iron, Lead, Mercury, Cadminium, Zinc, Copper, Nickel and Arsenic. They are varied from the quantities of these metals may be caused a deleterious causes, and plants and animals vary in this respect. For e. i.: 0.3mg/l of Zinc, 0.02 mg/l of Copper and 0.33mg/l of cause are lethal to sickle backs. Plants growth is inhibited by Zinc concentrations of 7mg/l or more, but 0.5mg/l of Copper or 0.01mg/l of Mercury will kill algae. One of the most important effects of metallic pollution is that aquatic organisms can absorb and accumulate concentrations of these metals in their tissues. Consequently increasing concentrations can bring up in food chains and actual, and they are more species of the secondary and tertiary tropic levels there may be up to 15 times as many Mercury, present in fish as in algae. Human populations are used the higher level consumers as food and there by ingest the metal toxins.

ii. **Pesticides Toxins:** Pesticides are a caused of water pollution. The most hazardous pesticides are the organ chloride compounds, because of their stability and persistence in the environment. Persistent organ chlorides may be accumulated in the food cycle e. g. Some shrimps and fish can concentrate some pesticides by a factor of 1000 to 10,000. Birds may have up to 55g/m3 of DDT in their fatty tissues.

iii. **Acid and Alkali Toxins:** Acids and alkalis may be contacted with hazardous because they decrease or increase the pH value of water from neutral value of pH 7. Most animals and plants will not survive in water with a pH value of below 5 (acid), or above pH value of 9 (alkaline). Changes in pH can be affected the actions of other toxins such as cyanides and sulphides, more toxic in acid conditions than they are in neutral or alkaline conditions.

iv. **Organic Compound Toxins:** Polychlorinated biphenyls are by-products of the plastic, lubricant, rubber and paper producing industries. Cyanides are very toxic to all biological life. Chlorophenols are toxic to bacteria and fish (1mg/l) and even to man (over 40mg/l).

4. Chemical nutrient effects

Chemical nutrient is very essential materials that are required by plants and animals for maintaining their growth and metabolism. Water pollution is the two most important nutrients they are Nitrogen and phosphorus, usually present in nitrates and phosphates form. Few amounts of nitrates and phosphates founds in all natural waters. The nutrient quantity slowly increases from the biodegradation of dead substances. This is increase in nutrients is called natural eutrophication. Increased concentrations of nitrates and phosphates in water produce the overall effect of increased rate of growth. A layer development on the surface of the water caused to as "bloom" condition. This blooms condition decrease the light penetration and restricts atmospheric supply oxygen of water. The effect is caused difficult conditions for river, canal and for swimming, bathing and fishing. The condense algae growth eventually dies, and the subsequent biodegradation produces an oxygen deficit which can result to noxious absence of O_2 conditions there is concern about increase nitrate concentration in drinking water absorbed from rivers, aquifers and boreholes. Nitrates are taken into the body in food and drink and can causes disease. This causes a reduction in the oxygen carrying capacity of the blood, and a condition called anoxia, or meth haemoglobin anaemia disease. This is particularly more young babies and can be fatal. It has been suggested that nitrites can be further converted to amines and nitrosamines in the human body, and these organic chemicals are a possible cause of gastric cancer.

5. Microorganism effects

Wastes are caused into water contain certain some pathogenic organisms that are capable of causing human diseases. Some bacteria cause; cholera, typhoid fever, bacillary dysentery and gastroenteritis. Viruses are also found in water, including strains which affect poliomyelitis, infectious hepatitis etc. The infective egg and larval stages of some animal parasites also found in water Ascaris lubricoides, (Round worm), the beef and pork tapeworm and Echinococcus Species (hydatid disease). All these types of organisms occur in faeces, and so are present in sewage and slurry.

6. Radionuclide effects

Development of nuclear energy is increase radioactive waste to dispose off into the environment, and it contains various radio nuclides with long half-lives. Example: Ruthenium-106, Strotinium- 90, Cerium – 144, Caesium – 137.

Conclusion

As the objective of the study to review of empirical evidence of the impact of pollutants of water. Water is a renewable natural resource. Due to ever increasing industrialization, urbanization, this precious resource is continuously under stress. There are multiple dimensions to water quality and its deterioration. Different types of water pollutants decrease the water quality make it public unacceptable which is called water pollution. Water is universal solvent of our body. Any living organism cannot live without water. Water pollution decreases the dissolve oxygen in water. Therefore highly polluted water bodies have more value of Biochemical Oxygen Demand (BOD). BOD measurement is a chemical method of water quality assessment. Water pollution affects the flora and fauna which are present on earth and in water bodies. Soil pollution and air pollution are major causes of water pollution. There are various new water purification techniques which have come up to purify water for example by using rechargeable polymer beads, using the seeds of *Moringa oleifera* tree, purifying water by using aerobic granular sludge technology etc. Research is being conducted all over the world to develop more and more techniques which can generate pure water at low cost. Therefore these results indicate that the use of multiple -markers sensitive to water pollution may provide complementary information to diagnose environmental factors that are impairing macro invertebrate communities. All these techniques are being developed to ensure that in near future.

References

Abdus-Salam, N. and Oyede, F.A., 2013. The Impact of a River on the Water Quality of Nearby Wells: A Case Study of Aluko River, Ilorin, Nigeria. ChemSearch Journal, 4(2), pp.10-20.

Ahuja, S. and Hristovski, K., 2013. Preface. In Novel Solutions to Water Pollution (pp. ix-x). American Chemical Society.

Almansa, C., J. Calatrava, J.M. Martínez-Paz, 2012. Extending the framework of the economic evaluation of erosion control actions in Mediterranean basins. Land Use Policy, 29(2): 294-308.

American Public Health Association, 1990. Water Pollution Control Federation. 1989. Standard methods for the examination of water and wastewater, 17.

American Public Health Association, American Water Works Association, Water Pollution Control Federation and Water Environment Federation, 1913. Standard Methods for the examination of water and wastewater. American Public Health Association.

American Public Health Association, American Water Works Association, Water Pollution Control Federation and Water Environment Federation, 1915. Standard methods for the examination of water and wastewater (Vol. 2). American Public Health Association.

Anampiu, J.M.M., 2011. Organochlorine pesticides residues in fish and sediment from Lake Naivasha (Doctoral dissertation, MSc thesis, University of Nairobi).

Anh, P.T., C. Kroeze, S.R. Bush, A.P.J. Mol, 2010. Water pollution by intensive brackish shrimp farming in south-east Vietnam: Causes and options for control. Agricultural Water Management, 97(6): 872-882.

Baccini, P., 2013. Lakes: chemistry, geology, physics. Springer Science & Business Media.

Bashar, A., Mazlin, M., Toriman, H., Barzani, M., Goh, C.T., Rahmah, E. and Rizal, M., 2013. The environmental risk and water pollution: A review from the river basins around the world.

Bhuiyan, A.B., Mokhtar, M.B., Toriman, M.E., Gasim, M.B., Ta, G.C., Elfithri, R. and Razman, M.R., 2013. The environmental risk and water pollution: A review from the river basins around the world. American-Eurasian Journal of Sustainable Agriculture, 7(2), pp.126-136.

Bockris, J.M., 1977. Environmental chemistry. In Environmental Chemistry (pp. 1-18). Springer US.

Cairns, J.E., 2017. Biological monitoring in water pollution. Elsevier.

Caporaso, J.G., Bittinger, K., Bushman, F.D., DeSantis, T.Z., Andersen, G.L. and Knight, R., 2009. PyNAST: a flexible tool for aligning sequences to a template alignment. Bioinformatics, 26(2), pp.266-267.

Chaudhry, F.N. and Malik, M.F., 2017. Factors Affecting Water Pollution: A Review. J Ecosyst Ecography, 7(225), p.2.

Cooper, P.F. and Findlater, B.C. eds., 2013. Constructed Wetlands in Water Pollution Control: Proceedings of the International Conference on the Use of Constructed Wetlands in Water Pollution Control, Held in Cambridge, UK, 24–28 September 1990. Elsevier.

Davies, C.M. and Bavor, H.J., 2000. The fate of stormwater-associated bacteria in constructed wetland and water pollution control pond systems. Journal of Applied Microbiology, 89(2), pp.349-360.

Gambhir, R.S., Kapoor, V., Nirola, A., Sohi, R. and Bansal, V., 2012. Water pollution: Impact of pollutants and new promising techniques in purification process. Journal of Human Ecology, 37(2), pp.103-109.

Greenstone, M. and Hanna, R., 2014. Environmental regulations, air and water pollution, and infant mortality in India. The American Economic Review, 104(10), pp.3038-3072.

Halder, J.N. and Islam, M.N., 2015. Water pollution and its impact on the human health. Journal of environment and human, 2(1), pp.36-46.

Harbaugh, A.W., Banta, E.R., Hill, M.C. and McDonald, M.G., 2000. MODFLOW-2000, The U. S. Geological Survey Modular Ground-Water Model-User Guide to Modularization Concepts and the Ground-Water Flow Process. Open-file Report. U. S. Geological Survey, (92), p.134.

Harter, T., 2003. Groundwater quality and groundwater pollution. UCANR Publications.

Hilsenhoff, W.L., 2017. An improved biotic index of organic stream pollution. The Great Lakes Entomologist, 20(1), p.7.

Hu, Y. and Cheng, H., 2013. Water pollution during China's industrial transition. Environmental Development, 8, pp.57-73.

Huang, F., Wang, X., Lou, L., Zhou, Z. and Wu, J., 2010. Spatial variation and source apportionment of water pollution in Qiantang River (China) using statistical techniques. Water research, 44(5), pp.1562-1572.

Kneese, A.V. and Schultze, C.L., 1975. Pollution, prices, and public policy: a study sponsored jointly by Resources for the Future, inc. and the Brookings Institution. The Brookings Institution.

Kneese, A.V., 2015. Water Pollution: Economics Aspects and Research Needs. Routledge.

Laws, E.A., 2017. Aquatic pollution: an introductory text. John Wiley & Sons.

Lu, Y., Song, S., Wang, R., Liu, Z., Meng, J., Sweetman, A.J., Jenkins, A., Ferrier, R.C., Li, H., Luo, W. and Wang, T., 2015. Impacts of soil and water pollution on food safety and health risks in China. Environment international, 77, pp.5-15.

Novotny, V., 1994. Water quality: prevention, identification and management of diffuse pollution. Van Nostrand-Reinhold Publishers.

Obasohan, E.E., Agbonlahor, D.E. and Obano, E.E., 2010. Water pollution: A review of microbial quality and health concerns of water, sediment and fish in the aquatic ecosystem. African journal of Biotechnology, 9(4).

Owa, F.W., 2014. Water pollution: sources, effects, control and management. International Letters of Natural Sciences, 3.

Panayotou, T., 2016. Economic growth and the environment. The environment in anthropology, pp.140-148.

Patterson, J.J., Smith, C. and Bellamy, J., 2013. Understanding enabling capacities for managing the 'wicked problem'of nonpoint source water pollution in catchments: a conceptual framework. Journal of environmental management, 128, pp.441-452.

Pintér, J.D., 2013. Global optimization in action: continuous and Lipschitz optimization: algorithms, implementations and applications (Vol. 6). Springer Science & Business Media.

Rickson, R.J., 2014. Can control of soil erosion mitigate water pollution by sediments?. Science of the Total Environment, 468, pp.1187-1197.

Rozell, D.J. and Reaven, S.J., 2012. Water pollution risk associated with natural gas extraction from the Marcellus Shale. Risk Analysis, 32(8), pp.1382-1393.

Sartor, J.D., Boyd, G.B. and Agardy, F.J., 1974. Water pollution aspects of street surface contaminants. Journal (Water Pollution Control Federation), pp.458-467.

Shiva, V., 2016. Water wars: Privatization, pollution, and profit. North Atlantic Books.

Shortle, J.S. and Dunn, J.W., 1986. The relative efficiency of agricultural source water pollution control policies. American Journal of Agricultural Economics, 68(3), pp.668-677.

Sikder, M.T., Kihara, Y., Yasuda, M., Mihara, Y., Tanaka, S., Odgerel, D., Mijiddorj, B., Syawal, S.M., Hosokawa, T., Saito, T. and Kurasaki, M., 2013. River water pollution in developed and developing countries: judge and assessment of physicochemical characteristics and selected dissolved metal concentration. CLEAN–Soil, Air, Water, 41(1), pp.60-68.

Stern, A.C. ed., 2014. Fundamentals of air pollution. Elsevier.

Van Donsel, D.J., Geldreich, E.E. and Clarke, N.A., 1967. Seasonal variations in survival of indicator bacteria in soil and their contribution to storm-water pollution. Applied microbiology, 15(6), pp.1362-1370.

Vaughan, W.J. and Russell, C.S., 2015. Freshwater recreational fishing: The national benefits of water pollution control. Routledge.

Verma, R. and Dwivedi, P., 2013. Heavy metal water pollution-A case study. Recent research in science and technology, 5(5).

Warren, C.E., 1971. Biology and water pollution control.

Wilbers, G.J., Becker, M., Sebesvari, Z. and Renaud, F.G., 2014. Spatial and temporal variability of surface water pollution in the Mekong Delta, Vietnam. Science of the Total Environment, 485, pp.653-665.

World Health Organization (WHO), 2017. Obesity and Overweight factsheet from the WHO. Health.

13

Air Pollution and Its Management

Any undesirable change in physical, chemical and biological properties of air is called air pollution. Air pollution interferes with wellbeing of living livelihood. Various human made and anthropogenic sources which deteriorate the air quality are called air pollutant. Some of the common air pollutant are dust, smoke, smog, fog, aerosol, SO_2, CO_2, O_3 and NO_x etc.

Sources of air pollution

The sources of air pollution are classified into two categories:

1. Natural source
2. Man-made source

1. Natural sources

- Ash from burning volcanoes, dust from storm, forest fires
- Pollen grains from flowers in air are natural sources of pollution

2. Anthropogenic (man-made) sources

They are population explosion, deforestation, urbanization and industrialization whose effect can be explained as fallow:

a) Power stations using coal or crude oil release CO_2 in air
b) Also furnaces using coal, cattle dung cakes, firewood, kerosene, etc.
c) Steam engines used in railways, steamers, motor vehicles, etc. give out CO_2.
d) So do Motors and internal combustion engines which run on petrol, diesel and kerosene. etc.
e) Vegetable oils, kerosene, and coal as household fuels
f) Sewers and domestic drains emanating foul gases
g) Pesticide residues in air
h) Carbon monoxide and Nicotine from smoking
i) Thermal power plants pollute air by emitted sulphur dioxide and fly ash
j) Fertilizers and pesticides

k) Mining activities release particulate matter into the air and pollutes
l) Sprayers
m) Burning of fuels

Major air pollutants

Air Pollutant: They are the substance which pollute the air. Some of the common pollutants are dust, soot, ash, CO, excess of CO_2, SO_2, CFC, Pb compounds, cement dust and radioactive rays.

They are two types and given below

1. Suspended particulate matter pollutants
2. Gaseous pollutants

1. Suspended particulate matter pollutants

- It is the dissolved solid or liquid matter in the atmosphere. Hence, also called as suspended particulate matter. These are mainly three types:
- **Dust:** It is suspended solid matter like cotton dust, iron dust etc. Their particle size is more than 1 µm and releases from the industrial chimney.
- **Aerosol:** These are dissolved solid or liquid particles released from jet planes, aerosols cans and sprays. They have a size of less than 1 µm.
- **Smog:** It is suspended liquid matter with size of more than 1 µm. Particles which are more than 10µm in diameter settle down.
- Major source of suspended particulate matter are vehicles, industries, thermal power plant, cement factories, refineries cotton factory.
- Fly ash and lead are major particulate air pollutant. Fly ash is the byproduct of thermal power plant or coal burning in chimney. In addition to air pollution, fly ash is also causes of water pollution. Direct deposition of fly ash on leaf surfaces affect the growth of plants. Fly ash also affect the human health and lung cancer. Fly ash is now being used for making bricks and as a land fill material.
- Lead is used in petrol or diesel to increase the octane number. Pb particles are released in air from vehicles by burning of petrol or diesel fuel. These particles have very bad effect on kidney and liver and also destroy the RBCs.
- During mining operation, various types of metal oxides like iron oxide, aluminium oxide, manganese oxide, zinc oxide etc. These particulate pollutant affect the physiological and biochemical activities of plants and animals.

2. Gaseous pollutants

Gaseous pollutants are different from particulate pollutant. Major gaseous pollutants are Carbon monoxide, carbon dioxide, sulphur dioxide, oxides of nitrogen. All of these have harmful effects on plants and humans.

- **CO:** This is most abundant air pollutant. CO is the result of incomplete combustion of fossils fuels. Diesel and petroleum are the source of carbon monoxide. CO is more dangerous than CO_2. It is a poisonous gas which causes respiratory problems. When it reaches the blood stream, it replaces oxygen due to its high affinity for hemoglobin.
- **CO_2:** Organic matter decomposition or burning of fossil fuels increase the CO_2 concentration in atmosphere. Its high concentration is act as pollutant. This is a more abundant green house gas which causes global warming. Higher level of CO_2 increases the atmospheric temperature. This is due to heat trapping ability of CO_2. Global warming is a major factor of climate change. High CO_2 concentration in aquatic ecosystem, form H_2CO_3 which increase the acidity Higher acidity in marine ecosystem decreases its productivity.
- **SO_2:** Thermal power plant, coal burning and petroleum refineries are major sources of SO_2. SO_2 damage, leather, paper and accelerates corrosion of metals, alloys and marbles. It is a phytotoxic pollutant and enhances chlorosis hence excessive sulphur dioxide pollution decreases crop yield. SO_2 also contribute to acid rains.
- **Oxides of nitrogen:** A few oxides of nitrogen, such as nitric oxide (NO), nitrous oxide (N_2O) and nitrogen dioxide (NO_2) are produced by natural processes as well as from thermal power stations, factories, automobiles and aircrafts (due to burning of coal and Health petroleum). They reduce the oxygen carrying capacity of blood, may cause eye irritation and skin cancer in human beings.
- Nitrogen oxides released into the atmosphere reacts immediately with gaseous hydrocarbons in the presence of sunlight to produce secondary pollutants like aldehydes, ozone and peroxy acetyl nitrate which accumulate as smoke.
- **Gaseous hydrocarbon and ozone:** These are the compounds of hydrogen and carbon released from automobile petroleum refineries and manufacturing of explosive. For example- Aniline damages the leaves of plant, 3,4- Benzopyrene causes lung cancer, Ethylene and acetylene cause neurosis and chlorosis of leaves.

Primary air pollutant and secondary air pollutant

- **Primary air pollutant:** These are release in atmosphere as native form as CO, DDT and NO_X.
- **Secondary air pollutant:** These are produced in environment due to reaction of primary pollutant. O_3, PAN, Aldehyde, Photochemical smog, London smog.
- **Smog:** Smog is a mixture of smoke, dust particles and small drops of fog. Smog may cause necrosis and develop a white coating on the leaves (silvering) of plants. In human beings and animals, it may cause asthma and allergies. Smog is measured by Ringleman method. Smog word is given by Desvoeux.

a) Lass Angels smog: First seen at Lass Angels place.

Light, O2, High Temp.

Smoke + NO_x + Fog ---------------------------> PAN + O_3 + NO_2

PAN and O_3 mixture is called Brown Haze. Brown haze formation is complete reaction of Nitrogrn oxides with fog and smoke. And Incomplete reaction generate gray haze.

- PAN stops the photolysis of H_2O during photosynthesis.
- PAN inhibit the PS-II.
- PAN inhibit the chlorophyll synthesis.
- This also affect the human health, causes irritation and lung cancer.

b) London smog (Sulphur smog): This was first seen at London

Low temp., moist atmosphere

Coal + SO_X + Fog -- H_2SO_4 Vapor

- London smog is the results of reaction of sulphur oxide with fog in the presence of low temperature and moist atmosphere. London smog is the sulphur smog.
- Due to entry of sulphur smog or H_2SO_4 vapour in the body by respiration process, there were dead 4000 people in London. So this called London smog.

Acid rain and its effect

- Acid rain word is given by Robert August
- No_2 and So_2 present in smoke and atmosphere water vapor, react to each other and formed (H_2So_4 + HNO_3) which comes on earth by rainfall, which is called Acid rain.

- Wet deposition- Acid on the earth comes through rain, fog and ice form, when it is called deposition.
- Dry deposition is defined as the acid, settled on the earth in the form of nitrate and sulphate.
- PH of acid rain is less than 5.6.
- In acid rain fall = 70% $H2SO_4$ and 30% HNO_3.

Effect of acid rain

- Acidity is increase in soil and water
- Acid rain is affected the historical buildings- Tajmahal, Redfort etc.

Control of air pollution

Different techniques are used for controlling air pollution caused by gaseous pollutants and that are caused by particulate pollutants.

Methods of controlling gaseous air pollutants

The gaseous pollutants like hydrocarbon, SO_2, NH_3 and CO etc. Controlled by using three different method

a) **Combustion.** This technique is applied when the controlling air pollutants in the form of organic gases or vapors. In this technique, the organic air pollutants are subjected to catalytic combustion technique or flam combustion. In this technique, organic pollutants are converted into less harmful products and water vapor.

b) **Absorption.** In this method substance penetrates into another substance like scrubbers. The gaseous pollutants are passed through absorbing material like scrubbers. These scrubbers contain a liquid absorbent. This liquid absorbent removes the pollutants present in gaseous effluents. Then the air coming into scrubber is free from pollutants and it is discharged into atmosphere.

c) **Adsorption.** Adsorption is a process in which a substance sticks to the surface of another substance also is called absorbent. In this technique, gaseous effluents are passed through porous solid absorbent kept in containers. The gaseous pollutants absorbed to the surface of the porous material and clean air passes through. The organic and inorganic constituents of gaseous effluents are trapped at the interface of solid adsorbent by physical adsorbent.

Methods to control particulate air pollutants

The air pollutants such as dust, soot, fly ash etc. cause the pollution. It can be controlled by using fabric filters, electrostatic precipitators, wet scrubbers and mechanical devices etc.

a) **Fabric filters:** In this technique, gaseous emission containing dust, soot and fly ash is passed through porous fabric filters made of fabric (cloth) (woven or filled fabric). The particles of pollutants get trapped in this fabric and are collected in the filter and the gases free from the pollutant particles are discharged.

b) **Mechanical devices:** There are many mechanical devices that clean the air of pollutants either due to gravity in which the particles settle down by gravitational force by sudden change in the direction of gas flow in which particles separate out due to greater momentum.

c) **Electrostatic precipitators:** Gases or air stream containing aerosols in the form of dust, mist or fumes, is passed between the two electrodes of the electrostatic precipitator. During this process, the aerosol particles get precipitated on the electrodes.

d) **Wet scrubbers:** They are use to trap SO_2, NH_3 and metals by passing fames through water.

Impact of air pollution on Environment

Air pollution can cause a variety of environmental effects: Acid rain is precipitation containing harmful amounts of nitric and sulfuric acids. These acids are formed primarily by nitrogen oxides and sulfur oxides released into the atmosphere when fossil fuels are burned. These acids fall to the Earth either as wet precipitation (rain, snow, or fog) or dry precipitation (gas and particulates). People are exposed to polluted air and their health and wellbeing are affected in socially unjust ways. Mitigating and overcoming this injustice does not mean equal exposure to environmental danger; it means elimination of every kind of pollution endangering the environment and living organisms. The challenge here lies in ensuring air quality, health and wellbeing for everyone. Health is not simplistically meant as absence of illness, but as a broader concept involving prevention as well as promoting physical, mental, and social wellbeing. It is a basic human value and a fundamental right of every individual regardless of race, religion, political beliefs, social, or economic status. Some are carried by the wind, sometimes hundreds of miles. In the environment, acid rain damages trees and causes soils and water bodies to acidify, making the water unsuitable for some fish and other wildlife. It also speeds the decay of buildings, statues, and sculptures that are part of our national heritage. Acid rain has damaged Massachusetts lakes, ponds, rivers, and soils, leading to damaged wildlife and forests.

1. Effects on wildlife

Toxic pollutants in the air, or deposited on soils or surface waters, can impact wildlife in a number of ways. Like humans, animals can experience health problems if they are exposed to sufficient concentrations of air toxics over time. Studies show that air toxics are contributing to birth defects, reproductive failure, and disease in animals. Persistent toxic air pollutants (those that break down slowly in the environment) are of particular concern in aquatic ecosystems. These pollutants accumulate in sediments and may biomagnify in tissues of animals at the top of the food chain to concentrations many times higher than in the water or air.

2. Ozone depletion

Ozone is a gas that occurs both at ground-level and in the Earth's upper atmosphere, known as the stratosphere. At ground level, ozone is a pollutant that can harm human health. In the stratosphere, however, ozone forms a layer that protects life on earth from the sun's harmful ultraviolet (UV) rays. But this "good" ozone is gradually being destroyed by man-made chemicals referred to as ozone-depleting substances, including chlorofluorocarbons, hydrochloro fluorocarbons, and halons. These substances were formerly used and sometimes still are used in coolants, foaming agents, fire extinguishers, solvents, pesticides, and aerosol propellants. Thinning of the protective ozone layer can cause increased amounts of UV radiation to reach the Earth, which can lead to more cases of skin cancer, cataracts, and impaired immune systems. UV can also damage sensitive crops, such as soybeans, and reduce crop yields.

3. Crop and forest damage

Air pollution can damage crops and trees in a variety of ways. Ground level ozone can lead to reductions in agricultural crop and commercial forest yields, reduced growth and survivability of tree seedlings, and increased plant susceptibility to disease, pests and other environmental stresses (such as harsh weather). As described above, crop and forest damage can also result from acid rain and from increased UV radiation caused by ozone depletion.

4. Global climate change

The Earth's atmosphere contains a delicate balance of naturally occurring gases that trap some of the sun's heat near the Earth's surface. This "greenhouse effect" keeps the Earth's temperature stable. Unfortunately, evidence is mounting that humans have disturbed this natural balance by producing large amounts of some of these greenhouse gases, including carbon dioxide and methane. As a result, the Earth's atmosphere appears to be trapping more of the sun's heat, causing the Earth's average temperature to rises a phenomenon known as global warming. Many scientists believe that global warming could have significant impacts on human health, agriculture, water resources, forests, wildlife, and coastal areas.

5. Effect of health

Air pollutants affect human health and pose significant threats to individuals worldwide, such as cardiovascular or respiratory disorders, asthma and lung cancer, which can be fatal. Given that an individual inhales more than 14.000 liters of air per day, it becomes obvious that air pollutants pose significant dangers for human health. According to the World Health Organization (WHO, 2006, 2009) more than two million premature deaths each year can be attributed to the effects of urban outdoor and indoor air pollution. In particular, every year indoor air pollution is responsible for the death of 1.6 million people while 800.000 deaths from lung cancer, cardiovascular and respiratory diseases worldwide are attributed to outdoor air pollution (Valent *et al.*, 2004; WHO, 2005). It is estimated that currently air pollution reduces average life expectancy of Europeans by 9 to 24 months (CEC, 2006). Do urban populations equally enjoy environmental quality? Intense social injustice is often located within cities and is related to the way in which the man-made environment is formulated, functions and develops in time and space. This evolution is a result of various social, cultural, economic, and political processes, along with environmental components and their interrelations, which combine to shape space consumption in an urban environment determining its residential quality from an environmental perspective.

Protecting the stratospheric ozone layers

- Ozone can be good or bad depending on where it is located.
- Close to the earth surface, ground level ozone is harmful air pollutant,
- Ozone in the stratosphere, high above the earth, protects human health and the environment from the sun's harmfull UV radiation.
- Ozone in the stratosphere, alayer of the atmosphere located 10 to 30 miles above the earth, serve as a shield, protecting people and the environment from the sun's harmful UV radiation. The stratospheric ozone layer filters out harmful sun rays, including a type of sunlight called UV. B Exposure to UV. B has been linked to cataracts (eye damage) and skin cancer. Scientist have also linked increased UV. B exposures to crop injury and damage to ocean plant life.
- Scientist became concerned that CFCs (chloro-fluro-carbon) could destroy stratospheric ozone. At that time CFCs were widely used

as aerosol propellants in consumer products such as hairsprays and deodorants and coolants in refrigerators and air conditioners.

- Montreal Protocol, 1987- Over 190 countries, including the major industrialized nations such as the united states, have signed the 1987 Montreal Protocol, which calls for elimination of chemicals that destroy stratosphere ozone.

Conclusion

In this chapter we attempted to present environmental quality in the context of environmental injustice. The discussion was based on data provided by international institutions mapping environmental degradation, its consequences on human health, wellbeing and life quality. These data reveal that environmental injustice within countries is particularly experienced by low socio-economic level groups of the population living in underprivileged areas. Moreover, environmental injustice between countries is more intense and degrades health quality of their population, indeed contributing to increase of mortality. the environmental problem of air pollution aiming at highlighting environmental injustice caused by the types, sources, and distribution of air pollutants along with their effects on human health and wellbeing. These issues were discussed in the scope of sustainability, which in our view is the only means to ensure environmental quality and justice for all. This is a requirement and a fundamental challenge for modern societies.

References

Cities. New England journal of medicine, 329: 1753–1759 (1993).

Dimitriou, A. and Christidou, V., 2007. Pupils' understanding of air pollution. Journal of Biological Education, 42(1), pp.24-29.

Dimitriou, A. and Christidou, V., 2011. Causes and Consequences of Air Pollution and Environmental Injustice as Critical Issues for Science and Environmental Education. In The Impact of Air Pollution on Health, Economy, Environment and Agricultural Sources. InTech.

Dockery, D. W. et al. An association between air pollution and mortality in six U.S.

Peters, J.M., Avol, E., Gauderman, W.J., Linn, W.S., Navidi, W., London, S.J., Margolis, H., Rappaport, E., Vora, H., Gong Jr, H. and Thomas, D.C., 1999. A study of twelve southern California communities with differing levels and types of air pollution: II. Effects on pulmonary function. American Journal of Respiratory and Critical Care Medicine, 159(3), pp.768-775.

Socha, T., 2007. Air Pollution Causes and Effects. Health and energy, 11.

Stockwell, W.R., Middleton, P., Chang, J.S. and Tang, X., 1990. The second generation regional acid deposition model chemical mechanism for regional air quality modeling. Journal of Geophysical Research: Atmospheres, 95(D10), pp.16343-16367.

Uriah, L.A. and Shehu, U., 2014. Environmental risk assessment of heavy metals content of municipal solid waste used as organic fertilizer in vegetable gardens on the Jos Plateau, Nigeria. American Journal of Environmental Protection, 3(6-2), pp.1-13.

Venkataraman, C., Sagar, A.D., Habib, G., Lam, N. and Smith, K.R., 2010. The Indian national initiative for advanced biomass cookstoves: the benefits of clean combustion. Energy for Sustainable Development, 14(2), pp.63-72.

14

Land Capability and Land Suitability Classification

Introduction

Land is one of the most valuable natural resources on Earth. However, its significance varies across professions. Engineers view land as a foundation for constructing roads, buildings, and flyovers, whereas agriculturists consider it a medium for cultivating crops. For business professionals, land represents a space for establishing offices and industries. Although "land" and "soil" are sometimes used interchangeably, they are distinct. Land is typically regarded as a two-dimensional surface, while soil is three-dimensional, extending in depth. Another key difference is that land refers to any portion of the Earth's surface, whereas soil consists of weathered rock materials enriched with essential nutrients, making it capable of supporting plant growth. The utilization of land should be aligned with its potential and limitations. Being a finite resource, land use is not solely determined by human needs but also by its capability. Over the past two decades, land degradation has become a growing concern globally, particularly in India. Land serves multiple purposes, primarily supporting agriculture, pastures, and forestry.

A standard soil survey map illustrates the various types of soil that are important and their distribution in relation to other landscape features. These maps are designed to cater to users with diverse needs and, therefore, contain detailed information to highlight essential soil distinctions. For a soil map to be useful, the information it presents must be clearly explained to the user. These explanations are referred to as interpretations. Soil maps can be understood in two ways: (1) by examining the individual soil types showed on the map and (2) by grouping soils that exhibit similar responses to management and treatment. Since there are numerous types of soil, there are also many specific soil interpretations. These interpretations provide users with all the relevant information that a soil map can offer. However, many users require more generalized information than that of individual soil-mapping units. To address this, soils are classified in different ways based on the specific needs of the map users. The composition of these soil groups and the allowable variation within each group depend on the intended use of the classification.

One of the key interpretive groupings is the capability classification, primarily developed for agricultural applications. Like other interpretive systems, this classification starts with individual soil-mapping units, which serve as the fundamental eletments of the system (table 1). In this classification, arable soils are categorized based on their potential and limitations for sustained cultivation of common crops that do not require special site modifications. Meanwhile, nonarable soils—those unsuitable for long-term cultivation—are grouped based on their ability to support permanent vegetation and the risks of degradation if mismanaged.

Soil maps indicate the location and distribution of different soil types. The most specific and accurate statements regarding soil use and management can be made about the individual mapping units represented on these maps. The capability classification of soils serves several purposes: (1) assisting landowners and other users in understanding and utilizing soil maps, (2) introducing users to the level of detail within a soil map, and (3) enabling broad generalizations regarding soil potential, limitations in use, and management challenges.

Land Capability Classification (LLC)

The classification of soils into capability classes and sub-classes is determined based on their potential to support crop and pasture growth sustainably over an extended period without degradation. This classification relies on evaluating the physical properties of the land and the extent of limitations affecting crop growth. The key factors considered include:

1. Inherent soil characteristics
2. External land features
3. Environmental constraints on land use

The first two aspects are derived from standard soil survey reports, while the third is obtained from additional sources.

The primary factors influencing soil capability are:

- Soil depth
- Drainage conditions
- Texture and structure
- Topography (slope)
- Erosion severity
- Susceptibility to flooding, overflow, and water saturation

- Presence of problematic soils, including salinity, alkalinity, acidity, and other adverse chemical conditions
- Climatic variability

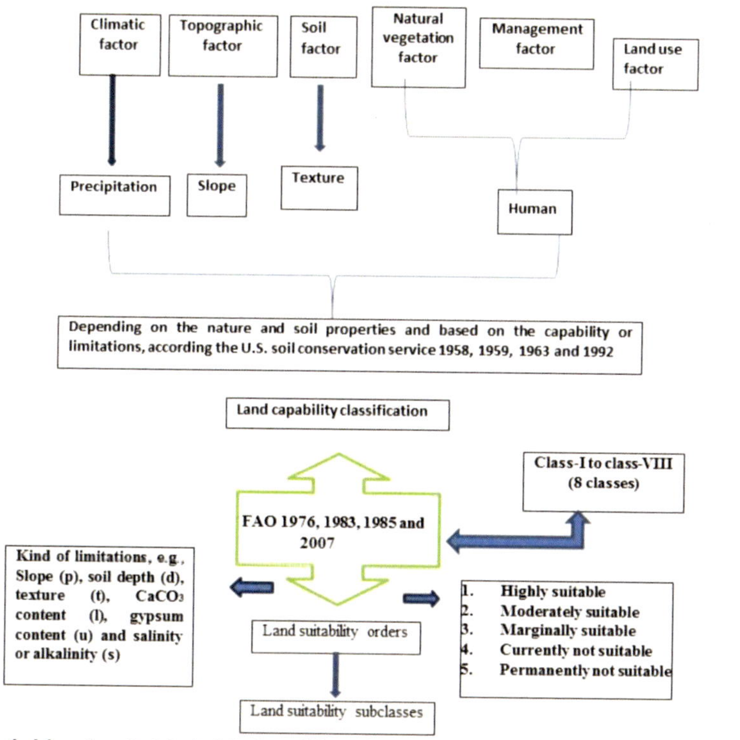

The methodology flow chart for both land capability and land evaluation classification (Mohamed *et al.* 2016)

The Land Capability Classification System was developed by the Soil Conservation Service of the USDA and is summarized in Table 9.1. This system consists of four major categories:

A. Land suitability class
B. Land capability classes (divided into eight classes)
C. Land capability sub-classes
D. Capability units

A. Land suitability class

Land suitability assessment is a specialized land evaluation approach introduced by the Food and Agriculture Organization (FAO, 1976) to determine the potential of a specific area for a particular crop rather than for general purposes. This analysis is essential for sustainable agriculture, as it evaluates how well a given land type aligns with a designated agricultural use. It involves assessing key factors such as climate, geography, and soil to determine land suitability for a specific crop. Among these, crop requirements and soil characteristics play a crucial role in evaluating the feasibility of cultivating a particular crop in a given area.

Land Suitability Classification

The **Land Suitability Classification** is a multi-tier system for evaluating soil, developed by the **FAO in 1976**. This classification follows a four-tier structure, consisting of **orders, classes, subclasses, and units**.

- At the **order level**, land is categorized into **suitable (S)** or **not suitable (NS)** based on its suitability and limitations for a specific use.
- Orders are further divided into **classes**, which reflect the degree of suitability and associated limitations.
- These **classes** are then subdivided into **subclasses**, which provide additional detail on specific limiting factors.
- Finally, **subclasses** are broken down into **land suitability units**, which define specific management requirements.

Orders and Classes of Land Suitability

Order S (Suitable)

1. **Class S1 (Highly Suitable)** – Land with no or minimal limitations for a particular sustainable use.
2. **Class S2 (Moderately Suitable)** – Land with moderate limitations that may affect sustainability.
3. **Class S3 (Marginally Suitable)** – Land with significant limitations that restrict its sustainable use.

Order NS (Not Suitable)

1. **Class N1 (Currently Not Suitable)** – Land with severe or very severe limitations that could potentially be improved over time, but not with current knowledge or at an acceptable cost.
2. **Class N2 (Permanently Not Suitable)** – Land with such extreme limitations that its use is not feasible under any foreseeable conditions.

B. Land capability classes

Soils are categorized into eight distinct capability classes based on their potential for agricultural use and the extent of limitations they possess. Class I soils exhibit the highest capability for agricultural management and have minimal limitations, making them the most productive and responsive to various management practices. In contrast, Class VIII soils have the lowest agricultural capability and face the most severe limitations, rendering them unsuitable for cultivation. Each capability class is closely associated with the cropping system, which is further influenced by the Length of Growing Period (LGP).

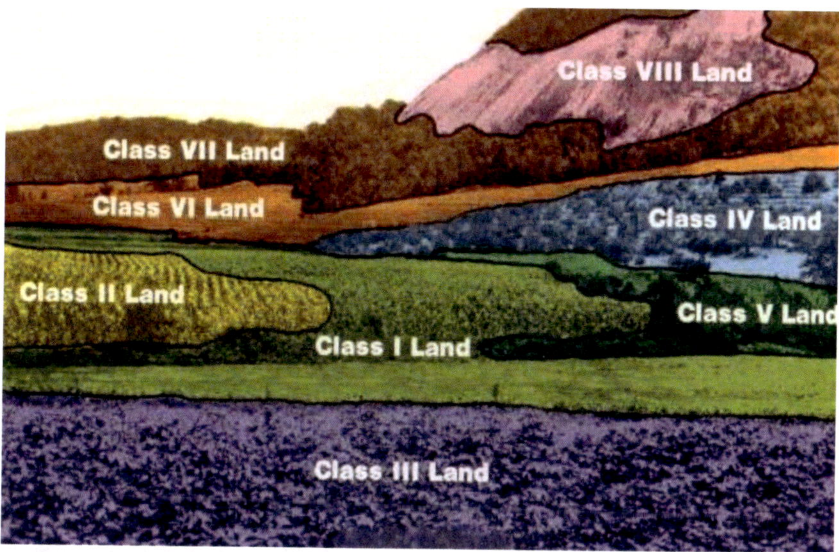

Table. 1. Land capability classifications (Eight classes)

Land suitable for cultivation				**Land not suitable for cultivation**			
Land capability class				**Land capability class**			
Class I Very good land (No limita-tion)	Class II Good land (Minor limita-tion)	Class III Moderately good (major limitation)	Class IV Fairly good (major limitation with occasional cultivation)	Suitable for pasture and grazing			Class VIII suitable for wild life and recreation
				Class V (No limitation)	Class VI (Minor limitation)	Class VII (Minor Limitation)	

Table. 2. Brief description / salient features of LCC

Name of the class	Colour shown on land capability	Salient features	Crop suitability crops	Management practices
Class – I	Green	Excellent cultivable deep, nearly level. Productive land with no limitation (or very slight hazard)	Suitable for most crops like wheat, barley, cotton, maize tomato and beat	Need no management practices for cultivation
Class – II	Yellow	Good cultivable land on almost level plain or on gentle slopes limitation of soil depth, salinity, texture, drainage/erosion that reduces the choice of plants	Suitable for wheat, barley, cotton	Cultivate with precaution need simple management practices.
Class III	Brown	Moderately good cultivable land on almost level plain or on moderate slope. Limitation :L Moderate erosion soil depth, soil salinity and soil texture. They may have vertic characteristics (or) drainage problem that reduces the choice of crops	Unsuitable for growing vegetable crops. Varying suitability for different crops	Cultivation with careful management practices need intensive care.
Class IV	Pink	Fairly good land on almost level plains or moderately steep slopes **Limitations** Strong / very strong soil salinity S3/S4 shallow soil depth, strong erosion, fine soil texture. Poor (or) excessive drainage.	Limited cultivation unsuitable for growing a variety of crops. Suitable for selected crops 8 for pasture. Not economical to cultivate	Need intensive soil conservation and management practices
Class V	Dark grey	**Limitations** Story, rocky nature (or) marshy-ness use of implements is difficult	Not suitable for arable farming suitable for grazing	

Class VI	Orange	Moderate limitation such as steep slope severe erosion, limiting soil depth, strongly gypsiferous, stony or sandy (sand – done areas) eg. dense forest lands of the Himalays. Sandy (gypsiferous) and donal areas of Thar Desert in Rajasthan and SW Irag.	Non-arable land well suited for grazing (or) forestry	
Class VII	Red	Severe limitation such as steeply sloping land subjected to erosion (or) very shallow stony soils that have not	Fairly well suited for grazing or forestry but not cultivable	Need careful management for grazing and forestry
Class VIII	Purple	Severe limitation Very steeply sloping, highly eroded, rocky barren mountain landscape, hyper arid rocky undulating surface	Non-arable, suitable for wildlife (or) recreation. Extremely rough, rocky, arid, wet (or) extremely saline land	

Table. 3. Criteria for Land Capability Classification (Gezachew and Ndao, 2008)

Parameters	Class I	Class II	Class III	Class IV	Class V	Class VI	Class VII	Class VIII
SLOPE (L)	0-8	8-16		16-32		30-50		>50
EROSION (e)	No sign to slightly	MODE RATE	High			Very High	Extremely High	
Stoniness (%)	0-40	>40				>40		
Soil Depth	>60	45-60	15-45	<15				
Soil Drainage	Never saturated	Rarely Saturated	Saturated for short period			Saturated for long period		
Soil Texture (t)	L, LS, SL	Si, SCL, SiCL, SiL,CL	SiC, SC	S, C				
Organic Carbon (%)	>1	0.8-1	0.6-0.8	0.4-0.6		0.2-0.4		
pH	5.5-7.9	4.5 to 5.5	<4.5	>8.4		<4.5->8.4		
Where, L-Loam, LS-Loamy Sand, SL-Sandy Loam, Si-Silty, SCL-Sandy Clay Loam, SiCL-Silty Clay Loam, SiL-Silty Loam, CL-Clay Loam, SiC-Silty Clay, SC-Sandy Clay, C-Clay, S-Sandy								

Capability Subclasses

Capability subclasses are determined based on the primary limiting factors that affect land productivity. These limitations include:

- **Wetness or Excess Water (w):** Indicates problems such as poor drainage or frequent waterlogging.
- **Climate (C):** Refers to constraints due to temperature extremes, inadequate rainfall, or other climatic factors.
- **Soil (S):** Represents soil-related limitations, including poor texture, low fertility, shallow depth, or salinity.
- **Erosion (e):** Denotes susceptibility to erosion, which may be caused by wind, water, or improper land use.

These subclasses are represented by appending limitation symbols to the capability class numbers. For example:

- **IIe** – Class II land with erosion as a primary limitation.
- **IIIw** – Class III land primarily affected by wetness or waterlogging.

Each subclass not only identifies the degree of limitation but also specifies the type of constraint that needs to be managed.

Notably, **Class I lands do not have subclasses**, as they are free from significant limitations and do not require special conservation measures.

D. Land Capability Units

Land capability units are further subdivisions of land capability subclasses. Soils within a single unit exhibit similar characteristics, making them suitable for the same crops and pasture plants. They require comparable management practices and demonstrate similar productivity and responses to agricultural practices. Each land capability unit shares common potential and limitations. These units serve as a practical guide for farmers in farm planning and land management. If existing constraints—such as inadequate irrigation, poor drainage, gully erosion, or flood risks—are mitigated through appropriate measures, the land's capability class can be improved. Conversely, a decline in conditions may downgrade its classification.

Derived from soil surveys, land classification provides a systematic approach to understanding soil properties. Recently, Geographic Information Systems (GIS) and Soil Information Systems (SIS) have enhanced the precision and usability of soil survey data. These technologies help map soil distribution, analyze soil variability, link data with spatial properties, and interpret land use information through comprehensive electronic databases. The adoption of GIS and SIS has significantly improved the efficiency and speed of soil surveys and land-use planning. These modern advancements became necessary due to the vast amount of soil and land survey data, which can now be systematically managed using advanced computer hardware and software.

Conclusion

Land capability and land suitability classification are essential tools in sustainable land-use planning and natural resource management. Land capability classification evaluates the inherent potential of land based on its physical and environmental constraints, such as soil type, slope, drainage, and erosion risk, categorizing land into classes that indicate its limitations for agricultural and other uses. On the other hand, land suitability classification is more specific, assessing how well a particular piece of land meets the requirements for a defined use, such as crop cultivation, forestry, urban development, or conservation. The integration of these classification systems ensures that land is utilized optimally while minimizing environmental degradation. By identifying limitations and opportunities, policymakers, farmers, and planners can make informed decisions that enhance productivity, preserve ecosystems, and support long-term sustainability. Advances in geospatial technologies, such as GIS and remote sensing, have further refined these classifications, allowing for more precise and dynamic assessments. Ultimately, land capability and suitability classifications serve as foundational frameworks for responsible land management, balancing economic development with ecological preservation. Their proper application can help mitigate land degradation, improve food security, and promote climate-resilient practices, ensuring that land resources remain viable for future generations. Continued research, updated methodologies, and stakeholder collaboration are crucial to adapting these systems to changing environmental and socio-economic conditions.

References

Kumar A (2019) "Soil Resource Inventory, Land Capability and Crop Suitability Assessment of Haradanahalli Micro-Watershed using Remote Sensing and GIS".Acta Scientific Agriculture 3(3): 129-137.

Anonymous (1961) Land-Capability Classification, Soil Conservation Service U.S. Department of agriculture. Agriculture handbook No. 210: 3-5.

FAO (1976) A framework for land evaluation. FAO soil bulletin no 32, Rome.

FAO (1983) Guidelines: land evaluation for rainfed agriculture, FAO soils bulletin no. 52, Rome.

FAO (1985) Guidelines: land evaluation for irrigated agriculture. Soils bulletin 55. Food and Agriculture Organization of the United Nations.

FAO (2007) A framework for land evaluation. FAO soil bulletin no. 6 Rome, Italy.

Ibrahim A (2016) A New approach to the land capability classification: Case study of Turkey, international Conference – Environment at a Crossroads: SMART approaches for a sustainable future. Procedia Environmental Sciences, 32:264-274. www.sciencedirect.com.

Mary TJ, Nowshaja PT (2016) Land Capability Classification of Ollukara Block Panchayat Using GIS, International Conference on Emerging Trends in Engineering, Science and Technology (ICETEST-2015). Procedia Technology, 24:303-308. www.sciencedirect.com.

Mohamed AE, Abdel R, NatarajanA, Hegde R (2016) Assessment of land suitability and capability by integrating remote sensing and GIS for agriculture in Chamarajanagar district, Karnataka, India.The Egyptian Journal of Remote Sensing and Space Sciences 19:125–141.

Panigrahy, S., Manjunath, K.R., Ray, S.S., 2006. Deriving cropping system performance indices using remote sensing data and GIS. Internationalournal of Remote Sensing, 26, 2595–2606.

Rajesh, N.L., U. Satishkumar, I. Shankergouda, S. N. Bhat, K. Basavaraj, H.V. Rudramurthy, K. Narayana Rao, V. Rajesh and Meenkshi Bai, R. 2018. Detailed LRI for Assessment of Land Capability and Land Suitability of Amarapura-2 Micro-Watershed Using RS and GIS. International Journal of Current Microbiology and Applied Science, 7(10):2370-2381. doi: https://doi.org/10.20546/ijcmas.2018.710.274.

Rao, D.P., Gautam, N.C., Nagaraja, R., Ram Mohan, P., 1996. IRSIC application in land use mapping and planning. Curr. Sci. 70, 575–578.

Sehgal J L (1996) Pedology - Concepts and applications.Kalyani Publishers, New Delhi.pp. 488.

Sehgal, J.L., 1996. Pedology-Concepts and Applications.Kalyani Publishers, Ludhiana.

Soil Conservation Service, 1963. Know the Capability of Your Land. USDA-NRCS, Washington, DC, pp. 1–25.

Soil Conservation Service, 1992, Washington, DC: Soil Conservation Service, pp. 60–73.

Soil Survey Staff, 1958, Land Capability Classification.Soils Memorandum SCS-22.

Soil Survey Staff, 1959, Land Capability Subclass.Soils Memorandum SCS- 30.Soil Conversation Service, Washington, DC.

15

Remote Sensing and GIS in Assessment, Monitoring and Management of Wastelands and Problem Soils

Introduction

Accurate and detailed information on the geographical location, aerial extent, and spatial distribution of wastelands is crucial for their efficient management and sustainable rehabilitation. Among modern resource assessment tools, Remote Sensing (RS) and Geographical Information Systems (GIS) have emerged as powerful technologies for detecting, evaluating, mapping, and monitoring degraded lands. Space-borne multispectral imagery, with its ability to provide large-scale, repetitive coverage, has proven particularly effective in identifying salt-affected soils and waterlogged areas in a timely and cost-efficient manner. In visual interpretation of satellite data, regions with high soil moisture and surface waterlogging appear as dark grey to black, indicating areas prone to water stagnation. To analyze the spatial dynamics of wastelands and assess the effectiveness of high-resolution satellite imagery, IRS P6 satellite data (23.5 m resolution) alongside topographical maps (1:50,000 scale) are commonly used. Given the dynamic nature of wastelands, multi-temporal satellite data is essential for precise delineation. These images must be geo-referenced within a unified coordinate system for consistency. The Planning Commission of India has endorsed GIS as an indispensable tool for land-use planning and wasteland development, facilitating treatment area identification, land-use trade-off analysis, and impact assessment simulations. Recognizing the importance of wasteland management in national development, the Ministry of Environment and Forests (MoEF) initiated a GIS-based wasteland mapping project in 1991, building upon earlier efforts by the Department of Science (1986) under the National Wasteland Identification Project, which mapped 147 districts at a 1:50,000 scale. A Task Force was established to develop a standardized classification system, later approved by the Planning Commission, categorizing wastelands into:

1. Gullied/Ravine Lands
2. Scrub-Uplands
3. Waterlogged/Marshy Areas
4. Saline/Alkaline Lands (Coastal/Inland)
5. Shifting Cultivation Zones
6. Sandy Deserts/Coastal Sands
7. Mining/Industrial Wastelands
8. Degraded Forest Lands
9. Overgrazed Pastures
10. Degraded Plantation Areas
11. Rocky/Stony Wastelands
12. Steep Sloping Terrain
13. Snow-Covered/Glacial Regions

The use of remote sensing data for mapping degraded lands began with the launch of the first Earth Resources Technology Satellite (ERTS-1/Landsat-1). However, with advancements in spatial and spectral resolution, satellites such as Landsat-TM, SPOT, and Indian Remote Sensing (IRS) satellites have significantly enhanced the efficiency of mapping and monitoring degraded lands. Geographic Information Systems (GIS) have proven to be powerful tools for managing spatial data at varying scales, processing extensive datasets—including soil characteristics, rainfall, temperature, and socioeconomic parameters—and conducting integrated analyses of regional resources. These capabilities aid in developing optimal solutions for various environmental and developmental challenges. Remote sensing imagery plays a crucial role in understanding land cover changes, making it an indispensable tool for tracking land degradation and desertification trends. Change detection, which involves identifying variations in an object's state over time, is vital for monitoring and managing natural resources and urban expansion. By providing a quantitative assessment of spatial distributions, such studies assist planners and stakeholders in promoting sustainable development. To analyze temporal changes in wastelands, multi-temporal datasets such as IRS Linear Imaging Self-Scanning (LISS III) are utilized. The geodatabase, created using ArcGIS and on-screen digitization techniques, reveals the type, extent, and spatial distribution of different wasteland categories. The process of wasteland mapping through satellite data involves data acquisition, preparatory work, and a well-defined methodology. Typically, this methodology includes base map preparation, visual interpretation of satellite imagery, development of classification legends, field data collection, soil sample analysis, categorization of degradation classes, and final map validation based on field observations and analytical data. For visual interpretation, geocoded False Colour Composite (FCC) products at a 1:50,000 scale are employed. Image attributes such as color, tone, texture, pattern, shape, size, location, and association help in identifying and delineating various types of wastelands. However, these delineations require field validation, as interpretation varies with season, scale, and image resolution. Some categories, like salt-affected lands, waterlogged or marshy areas, and sandy regions, can be distinctly identified due to their unique spectral characteristics and patterns. In contrast, gullied or ravine lands and shifting cultivation areas require moderate effort to delineate, while undulating uplands with or without scrub pose challenges due to their spectral similarity to fallow lands. Multi-date imagery can help address such complexities. Using multi-spectral satellite data, maps depicting the extent, spatial distribution, and severity of erosion, salt-affected soils, waterlogging, and shifting cultivation have been generated at scales of 1:250,000 and 1:50,000. Saline soils appear in varying shades of white with fine to coarse textures on FCC satellite imagery, making them distinguishable under normal crop growth conditions. To assess these soils, the National Remote Sensing Agency (NRSA), in collaboration with central and state government organizations, has produced maps at a 1:250,000 scale using Landsat Thematic Mapper (TM) and IRS satellite data. A digital atlas for India has also been developed, supporting land reclamation and soil conservation initiatives. Currently, the estimated wasteland area in India is approximately 63.85 million hectares. Each map provides details at the village, forest compartment, and micro-watershed levels, facilitating targeted planning and resource management.

Remote Sensing and GIS

Remote Sensing is the science of acquiring information about objects or phenomena by detecting emitted or reflected electromagnetic radiation (EMR). The most common form of remote sensing involves passive systems, which record naturally occurring EMR, such as visible light (blue, green, red) and near-infrared reflectance, as well as thermal infrared energy emitted from the Earth's surface. In contrast, active remote sensing systems do not depend on external radiation sources like the Sun or Earth's thermal emissions. Instead, they generate their own signals (e.g., radar or LiDAR) to measure responses from the target.

Remote sensing is both an art and a science, utilizing sensors aboard satellites or aircraft to collect Earth observation data. These sensors capture imagery, which is then processed, analyzed, and visualized using specialized techniques.

The resulting remotely sensed data is often integrated into a Geographic Information System (GIS) for further spatial analysis.

GIS (Geographic Information System) is a computer-based tool for mapping and analyzing geospatial data. It combines traditional database functions—such as querying and statistical analysis—with dynamic mapping capabilities. GIS organizes location-based information, enabling users to visualize and interpret trends in demographics, economic development, land use, and environmental factors. By linking databases with maps, GIS facilitates interactive data exploration, overlay analysis, and advanced modeling—capabilities that surpass traditional spreadsheets. These features make GIS indispensable for decision-making, forecasting, and strategic planning across diverse industries, including urban planning, agriculture, disaster management, and business logistics.

Process of Remote Sensing

Remote sensing involves detecting and recording the radiant energy reflected or emitted by objects or surface materials. It is a method of gathering information about the Earth's surface without direct physical contact. This process relies on capturing and analyzing electromagnetic radiation, followed by data processing and application. The interaction between incident radiation and the target objects plays a crucial role in remote sensing.

Key Stages in Remote Sensing

a) **Emission of Electromagnetic Radiation (Energy Source):** The process begins with a source of electromagnetic energy that illuminates the target. This could be natural (such as the Sun) or artificial (such as a radar system).

b) **Radiation Interaction with the Atmosphere:** As electromagnetic energy travels from the source to the target, it passes through the atmosphere, where it may be absorbed, scattered, or altered. A second interaction with the atmosphere may occur as the energy travels back from the target to the sensor.

c) **nteraction with the Target:** Upon reaching the target, the radiation interacts with it based on the material properties of the object and the characteristics of the energy itself. This interaction determines the type and amount of energy that will be reflected, absorbed, or transmitted.

d) **Recording of Energy by the Sensor:** A sensor, positioned remotely (not in direct contact with the target), captures and records the reflected or emitted electromagnetic energy. This data serves as the foundation for further analysis.

e) **Transmission, Reception, and Processing:** The recorded data is transmitted—typically in electronic form—to a receiving station. Here, it undergoes processing to convert the raw data into a usable format, such as an image, either in digital or hardcopy form.

f) **nterpretation and Analysis:** The processed images are analyzed either visually, digitally, or electronically to extract valuable information about the target. This step helps in identifying patterns, characteristics, or changes in the observed area.

i) **Application of Remote Sensing Data:** The final stage involves applying the extracted information to various fields such as environmental monitoring, agriculture, disaster management, land use planning, and more. This helps in gaining deeper insights, discovering new patterns, or solving specific problems.

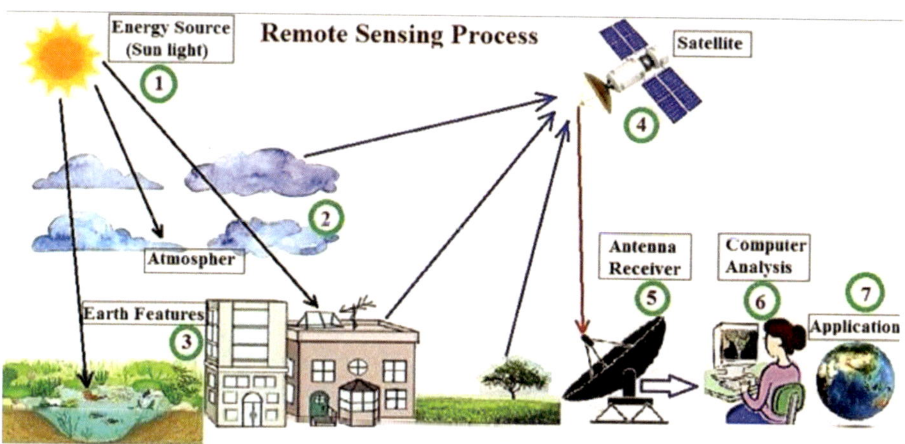

Fig. 1. Component of remote sensing

Electromagnetic Energy

Electromagnetic energy propagates in waves and spans a broad spectrum of wavelengths, from long radio waves to short gamma rays (Fig. 2). Only a small portion of this spectrum—visible light—is detectable by the human eye. Different technologies utilize distinct segments of the spectrum; for example, radios operate within one range, while X-ray machines employ another. NASA's scientific instruments study Earth, the solar system, and the wider universe by harnessing the full electromagnetic spectrum.

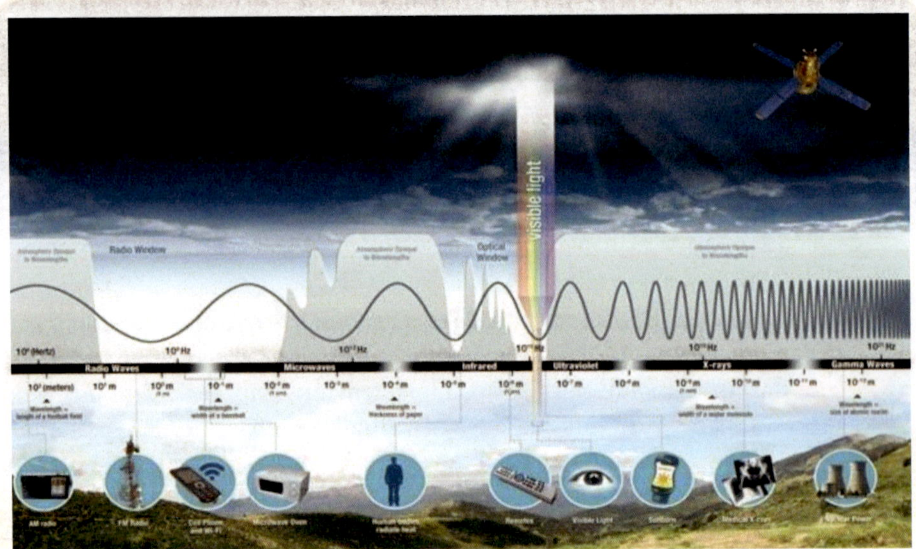

Fig 2: Electromagnetic Spectrum

Image Interpretation Keys – Color/Tone

- **White**: Represents objects with high light reflectance – **Saline soil, sandy soil**.
- **Black**: Represents objects with low light reflectance – **Black soil, fire-affected forest areas**.
- **Green**: Indicates objects with a reddish nature – **Red soil**.
- **Red**: Indicates objects with a greenish nature – **Agricultural land, grasslands, forests**.
- **Blue**: Represents objects with a bluish nature – **Water bodies**.

Land degradation assessment through Remote Sensing and GIS

Land degradation assessment using remote sensing relies on the spectral reflectance of surface soils. Degraded soils exhibit distinct spectral signatures that differ from those of healthy soils, enabling their identification through remote sensing techniques.

- Eroded soils show higher reflectance in the visible and near-infrared (NIR) regions.
- Saline soils appear bright white in summer but lose this reflectance during the monsoon season.
- Acidic soils display a slight yellowish-orange hue due to higher iron content.

- Surface ponding can be detected by a persistent blue tone in multi-seasonal imagery.

Mining areas may also appear blue in monsoon images (similar to ponds) because of water stagnation, which can be verified using landform and ancillary data. However, sodic and calcareous soils lack distinct spectral reflectance, making them difficult to delineate through visual interpretation alone. These soils are typically mapped using historical soil data and ground truth verification. Additionally, factors such as vegetation cover, land management practices, and tillage operations can alter soil reflectance, complicating the precise mapping of degraded land.

Utilizing Remote Sensing for Problematic Soils

In agriculture, fertile soil is crucial for successful crop production. However, not all soils are equally suitable, and problematic soils can pose significant challenges. Fortunately, remote sensing technology has emerged as a transformative solution for addressing these soil-related issues.

1. **Erosion Risk Assessment:** Soil quality and land stability are persistently threatened by erosion. Remote sensing enables comprehensive erosion risk assessment by providing critical insights into erosion patterns through satellite imagery and aerial surveys. By identifying vulnerable areas and understanding erosion mechanisms, targeted mitigation strategies can be implemented to protect topsoil and prevent land degradation.
2. **Irrigation Efficiency:** Efficient water management is vital in agriculture, especially in water-scarce regions. Remote sensing tools allow precise monitoring of soil moisture levels, facilitating optimized irrigation practices. With real-time data, farmers can reduce water waste, lower operational costs, and enhance crop productivity by ensuring optimal water usage.
3. **Erosion Mapping:** specialized application of remote sensing, erosion mapping offers detailed insights into erosion dynamics. It involves creating erosion risk maps that highlight areas susceptible to soil loss. These maps serve as vital tools for land-use planning and conservation strategies. By pinpointing high-risk zones, land managers can implement targeted erosion control measures. Explore its critical role in geomorphology.

4 **Vegetation Cover Assessment:** Healthy vegetation is key to preventing soil erosion and maintaining soil fertility. Remote sensing simplifies vegetation cover assessment by enabling the analysis of satellite imagery. This allows for tracking changes in vegetation density and

health over time. Such data supports sustainable land management and helps evaluate the success of reforestation efforts.

5. **Soil Moisture Monitoring:** Regular monitoring of soil moisture is essential for agriculture and natural resource management. Remote sensing provides a comprehensive, non-invasive method to track soil moisture variations. Farmers and environmentalists can leverage this data to optimize planting, irrigation, and conservation strategies. Understand its significance in hydrology.
6. **Flood Risk Assessment:** In flood-prone regions, remote sensing plays a crucial role in risk evaluation. By analyzing topographic data and historical flood patterns, experts can improve flood prediction and mitigation. This technology aids communities in developing flood-resilient infrastructure and early warning systems. Learn more about its applications in flood management.
7. **Soil Degradation Analysis:** Soil degradation, often driven by human activities, poses a major threat to agricultural sustainability. Remote sensing facilitates the assessment of degradation extent and causes. Using satellite imagery and soil quality data, researchers can identify degraded areas and devise restoration plans.
8. **Tracking Soil Erosion:** Preserving soil quality necessitates continuous monitoring of erosion. Remote sensing provides an efficient and cost-effective method to assess erosion rates over time. By analyzing data from different periods, we can evaluate the effectiveness of erosion control strategies and implement necessary adjustments. Additionally, it plays a crucial role in understanding the impact of deforestation on soil erosion.
9. **Soil Nutrient Management:** Maintaining a balanced nutrient composition in soil is essential for optimal crop growth. Remote sensing technology aids in nutrient management by providing insights into soil nutrient levels. This data allows farmers to optimize fertilizer application, reducing waste and minimizing environmental impact.
10. **Identifying and Managing Soil Salinity:** Excessive soil salinity can degrade ecosystems and render land unproductive. Remote sensing enables the creation of detailed salinity maps, highlighting areas with high salt concentrations. With this information, effective salinity management strategies can be implemented, such as selecting salt-tolerant crops and focusing on land reclamation efforts.

Using Remote Sensing to Map Soil Salinity

Soil salinization is driven by environmental factors such as soil type, land use, climate, and topography, as well as anthropogenic influences like poor drainage systems and excessive irrigation. The availability of airborne and spaceborne remote sensing platforms has significantly enhanced environmental monitoring by providing vast datasets applicable to risk mitigation, land surveys, sustainable agriculture, and climate change studies. These systems offer improved spatiotemporal and spectral resolutions, enabling researchers to track inland changes and identify salinity patterns across different geographical scales. Additionally, ground-based tools like electromagnetic induction (EMI) instruments, when combined with geospatial data, provide valuable insights into salinization at local and canopy levels, helping policymakers better understand field-scale dynamics. Studies have demonstrated that integrating remote sensing data with robust analytical methods holds great potential for salinity mapping. For instance, research in Iraq's Qom Valley successfully predicted soil salinity by combining Landsat 8 OLI spectral indices with topographic features. Similarly, in China's northern Tarim Basin, a machine learning model incorporating Sentinel-2 MSI data and laboratory measurements provided a reliable reference for assessing salinization in arid regions. Other studies, such as those in central Iraq using ALOS/PALSAR radar data and Landsat 5 TM, have shown cost-effective alternatives for salinity monitoring.

Further supporting this approach, researchers mapped salt distribution on the Great Hungarian Plain using Landsat 8 OLI-derived spectral indices and principal components in a multiple linear regression model, achieving 75% accuracy and outperforming ridge regression. These findings highlight how remote sensing variables can enhance localized soil salinity assessments. Since land cover and land use changes correlate with salinization extent, remote sensing data can efficiently estimate soil salinity levels, offering a powerful tool for sustainable land management.

Conclusion

Remote Sensing (RS) and Geographic Information Systems (GIS) have revolutionized the assessment, monitoring, and management of wastelands and problem soils by providing accurate, cost-effective, and real-time spatial data. These technologies enable the systematic identification and classification of degraded lands through multispectral and hyperspectral imagery, allowing for the detection of soil salinity, erosion, waterlogging, and other forms of land degradation. GIS facilitates the integration of various datasets—such as soil type, land use, climate, and topography—to generate detailed maps and models that support decision-making. By employing time-series satellite data (e.g., Landsat, Sentinel, and MODIS), changes in land degradation can be monitored over time, helping policymakers and land managers implement targeted rehabilitation strategies such as afforestation, soil amendments, and sustainable agricultural practices. Additionally, machine learning and AI-driven analysis enhance the predictive capabilities of these tools, enabling early warning systems for further degradation. The combined use of RS and GIS ensures efficient land resource management, aiding in the restoration of wastelands and improving soil health for agricultural productivity and ecological balance. Governments and environmental agencies must continue to invest in these technologies to combat land degradation, achieve sustainable development goals, and ensure food security for future generations. Thus, RS and GIS are indispensable tools in the fight against land degradation, offering scalable, precise, and dynamic solutions for the sustainable management of wastelands and problem soils worldwide.

References

Abdel Rahman, M. A., Natarajan, A., Srinivasamurthy, C. A., & Hegde, R. (2016). Estimating soil fertility status in physically degraded land using GIS and remote sensing techniques in Chamarajanagar district, Karnataka, India. The Egyptian Journal of Remote Sensing and Space Science, 19(1), 95-108.

Anthony, T. (2023). Assessment of Heavy Metal Contamination in Wetlands Soils Around an Industrial Area Using Combined GIS-Based Pollution Indices and Remote Sensing Techniques. Air, Soil and Water Research, 16, 11786221231214062.

Barrett, B. W., Dwyer, E., & Whelan, P. (2009). Soil moisture retrieval from active spaceborne microwave observations: An evaluation of current techniques. Remote Sensing, 1(3), 210-242.

Masoud, A. A., Koike, K., Atwia, M. G., El-Horiny, M. M., & Gemail, K. S. (2019). Mapping soil salinity using spectral mixture analysis of landsat 8 OLI images to identify factors influencing salinization in an arid region. International Journal of Applied Earth Observation and Geoinformation, 83, 101944.

Mensah, F., Adanu, S. K., & Adanu, D. K. (2015). Remote sensing and GIS based assessment of land degradation and implications for Ghana's ecological zones. Environmental Practice, 17(1), 3-15.

Pandey, A., Chowdary, V. M., Mal, B. C., & Dabral, P. P. (2011). Remote sensing and GIS for identification of suitable sites for soil and water conservation structures. Land Degradation & Development, 22(3), 359-372.

Reddy, G. O., Kumar, N., & Singh, S. K. (2018). Remote sensing and GIS in mapping and monitoring of land degradation. Geospatial technologies in land resources mapping, monitoring and management, 401-424.

Singh, G., Bundela, D. S., Sethi, M., Lal, K., & Kamra, S. K. (2010). Remote Sensing and Geographic Information System for Appraisal of Salt-Affected Soils in India. Journal of environmental quality, 39(1), 5-15.

Trivedi, A., & Awasthi, M. K. (2021). Runoff estimation by Integration of GIS and SCS-CN method for Kanari River watershed. Indian Journal of Ecology, 48(6), 1635-1640.

Trivedi, A., & Gautam, V. K. (2022). Decadal analysis of water level fluctuation using GIS in Jabalpur district of Madhya Pradesh. Journal of Soil and Water Conservation, 21(3), 250-259.

Venkataratnam, L., & Sankar, T. (1996). Remote sensing and GIS for assessment, monitoring, and management of degraded lands. Surveillance des Sols dans l'Environment par Télédétection et Systèmes d'Information Géographiques (Monitoring Soils in the Environment with Remote Sensing and GIS), 503-516.

Vikram, L., & Bhardwaj, M. (2023). Assessment of Urban Wastelands using GIS and IoT as Tools for Spatial Data Analysis. Current World Environment, 18(2), 893.

16

Multipurpose Trees: Their Selection and Role in Land-Use Systems

Introduction

Agroforestry, social forestry, community forestry, village forestry, and farm forestry are all practices that involve tree cultivation primarily outside designated forest reserves. Agroforestry, in particular, is a land-use approach that integrates trees with agricultural crops, shrubs, pastures, or livestock. This combination of trees and shrubs within land management systems can be structured in a spatial arrangement or occur in a sequential manner over time. The term 'woody perennials' is sometimes used to describe these trees and shrubs, referring to all long-lived plants that persist for more than a year, including bamboos and palms. In agroforestry systems, trees may not always be deliberately planted; natural regeneration may be encouraged by protecting emerging seedlings, or mature trees may be intentionally retained in fields and pastures. These woody perennials are often categorized as "multipurpose trees" (MPTs) or "multipurpose trees and shrubs" (MPTS) due to their diverse uses. While most trees can be considered multipurpose, this distinction highlights the varied benefits trees offer in agroforestry systems, as opposed to single-purpose timber plantations focused solely on wood production.

Conversely, tree cultivation in designated forest areas is often directed toward industrial needs, commonly referred to as industrial forestry. The promotion of MPTs in agroforestry holds great promise for addressing land-use challenges and improving livelihoods by enhancing environmental sustainability and economic opportunities for local communities. Agroforestry research has now advanced to a stage where suitable tree-crop combinations, their management practices, and genetically improved planting materials can be recommended for specific conditions. Furthermore, continuous advancements in research by international and national agroforestry organizations are contributing to the benefit of farmers. The availability of information on multipurpose tree species (MPTs) for many species has facilitated the adoption of agroforestry as a viable land-use system. However, for certain MPTs, a lack of fundamental data on growth rates, breeding patterns, resilience to environmental stress, and response to management still poses challenges. Selecting MPTs that are well-suited to agroecosystems and fulfilling the needs of local communities or industrial requirements will further promote agroforestry adoption. Additionally, the development of genetically improved high-yielding clones within these tree species, designed to meet specific agroforestry objectives, will enhance the appeal and acceptance of agroforestry practices among farmers.

What is a Multipurpose Tree?

Multipurpose trees (MPTs) play a vital role in agroforestry systems. These are woody perennials intentionally planted or retained within a land-use system to provide multiple products and benefits. Various definitions of MPTs exist, with Burley and von Carlowitz (1984) offering a synthesized definition:

"Multipurpose trees and shrubs (MPTS) are those that are deliberately grown, retained, and managed for more than one intended use, typically for significant economic and/or ecological benefits within multipurpose land-use systems, especially agroforestry."

Similarly, the Forestry/Fuelwood Research Development Project (1992) defined MPTs as:

"Tree species cultivated to deliver more than one major crop or function on a farm. On small farms, this often means utilizing both the wood and leaves from the same tree."

A wide range of tree species falls under the category of MPTs, spanning diverse taxonomic groups. Farmers have been cultivating trees for various purposes for thousands of years. In essence, all trees provide shade and help in preventing soil erosion, which means they serve at least two functions. However, the term "multipurpose tree species" specifically refers to trees grown for multiple essential uses, such as fuelwood, timber, fiber, fodder, food, medicine, and environmental services like soil conservation, fertility improvement, climate regulation, flood control, and carbon sequestration.

Trees can be classified as multipurpose in two ways:

1. **A single tree species can yield multiple products:** For example, in Central America, farmers grow *Gliricidia sepium* as a living fence, providing fuelwood, fodder, and green manure simultaneously. Similarly, in tropical regions, jackfruit (*Artocarpus heterophyllus*) serves as a source of food, fodder, and timber.

2. **The same species can be managed differently to yield different products.** For instance, in tropical regions, *Leucaena leucocephala* is cultivated such that some trees primarily produce wood, while others are managed to yield leaf fodder.

Thus, multipurpose trees contribute significantly to sustainable farming and environmental conservation by fulfilling multiple roles within a single land-use system.

Why Grow Multipurpose Trees?

Mitigating the Risk of Crop Failure

Cultivating multipurpose trees helps safeguard against total crop failure. For instance, if farmers rely on Leucaena leucocephala for livestock fodder and its foliage is damaged by pests, the remaining wood can still be used for fuel, pulp, or lightweight construction. Agroforestry systems enhance farm resilience by promoting plant diversity, reducing dependency on a single crop. This diversification acts as a natural insurance policy—if one crop suffers due to pests or market fluctuations, others can compensate. Additionally, a mix of species on a farm lowers vulnerability to specific pests.

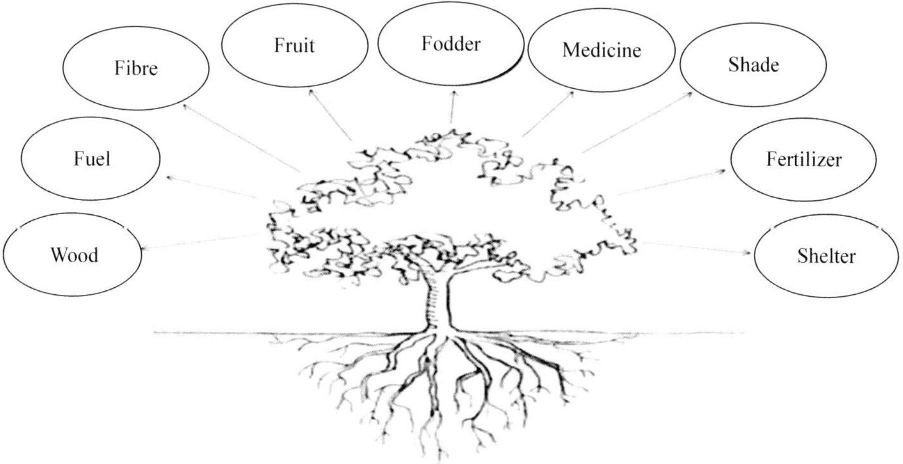

Enhanced Income Generation and Stability

Agroforestry practices contribute to stable and increased farm income. Different tree species provide harvests at various times of the year, ensuring a steady revenue stream. For example, in many Asian countries, farmers use the leaves of Artocarpus heterophyllus (jackfruit) as livestock feed during dry seasons, while the fruit serves as a source of income and food. Over time, the mature tree provides valuable timber. This system helps distribute income and employment opportunities more evenly throughout the year (Alam, 1993).

Diverse Production Benefits

Woody perennials offer a wide array of products and advantages, recognized by both researchers and practitioners. Initially, agroforestry's value was highlighted by listing multiple uses for trees and shrubs. While this principle remains valid, it is now understood that categorizing these benefits improves clarity. Trees contribute to both production and service functions, with some directly supporting agricultural productivity. Multipurpose trees (MPTs) on farmlands provide significant economic benefits at individual, community, and national levels by yielding food, fuelwood, fodder, organic fertilizers, timber, medicinal resources, and more.

Crucial Ecosystem Sertvices

Integrating trees into agroforestry systems enhances ecological balance. Trees help regulate soil temperature by providing shade, reduce soil moisture loss, and prevent erosion caused by wind and water. Additional ecosystem benefits include carbon sequestration, climate change mitigation, nitrogen fixation, improved soil fertility, land stabilization, biodiversity conservation, flood control, air and water purification, pest regulation, and enhanced aesthetic value. These ecosystem services contribute to a more sustainable and resilient agricultural landscape.

MPT attributes in relation to production and service functions	
Attributes	**Effects**
Breeding pattern, outcrossing or inbreeding, pollination method	Related to production and service functions; variation found in populations of seedling origin
Dioecious or monoecious	Distribution of sexes within and between individual plants: important for seed and fruit production and pollen flow
Tree height	Ease of harvesting leaf, fruit, seed, branch wood; shading effects
Stem form	Suitability for timber, posts, poles; shading effects
Crown size and form	Quantity of leaf, mulch and fruit production; shading effects
Multi-stemmed habit	Fuelwood production; shading effects
Rooting pattern (deep or shallow; spreading or geotrophic)	Competitiveness with other components, particularly resource sharing with crops; suitability for soil conservation
Leafiness; physical and chemical composition of leaves and pods	Fodder and mulch yield and quality; soil nutritional aspects
Thorniness	Suitability for live fencing or hedgerow intercropping
Wood quality	Acceptability for fuelwood and various wood products
Phenology: leaf flush, leaf fall, flowering and fruiting cycle; seasonality	Timing and labour demand for fruit, fodder, seed harvest; ability to withstand extreme conditions

Deciduousness	Seasonal or permanent leaf fodder availability; suitability for live fences, hedges and shelterbelts
Pest resistance; vigour	Major requirements, irrespective of function
Site adaptability and ecological range	Suitability for extreme sites or reclamation uses
Response to pruning and cutting	Use in hedgerow intercropping or for pollarding, lopping, coppicing
Possibility of nitrogen fixation	Use in hedgerow intercropping, planted fallows, rotational systems

Bioremediation and Agroforestry: An Integrated Approach

Definition and Importance of Bioremediation Bioremediation is the process of using biological organisms to mitigate environmental issues such as soil or groundwater contamination. Similarly, agroforestry refers to land-use systems where trees, agricultural crops, and/or livestock coexist in a structured spatial or temporal arrangement, ensuring ecological and economic interactions. This system effectively utilizes marginal and degraded lands by improving soil quality through tree-based interventions.

Key Aspects of Sustainable Land Use and Reclamation

1. Enhancing the sustainability of marginal lands through reclamation and restoration.
2. Developing plant varieties suited to challenging soil conditions.
3. Maintaining soil organic matter and biological activity to enhance soil physical properties and nutrient balance.
4. Improving nutrient cycling and efficiency in agro-ecosystems.
5. Utilizing fertilizers and other inputs strategically to address deficiencies.
6. Enhancing water-use efficiency.

Role of Trees in Soil Improvement Trees play a crucial role in maintaining and enhancing soil fertility. This is evident in the nutrient-rich conditions of natural forests, the soil restoration achieved through shifting cultivation, and the contributions of agroforestry to land reclamation. Soil surveys often reveal higher organic matter and improved physical properties beneath tree canopies. Notably, species such as *Faidherbia albida* have been shown to enhance crop yields under their canopy.

Mechanisms by Which Trees Improve Soil Fertility

1. Increasing organic matter input to the soil.
2. Reducing soil nutrient losses.
3. Enhancing soil physical, chemical, and biological properties.

4. Preventing runoff and soil erosion.
5. Maintaining soil organic matter and improving structure.
6. Increasing nutrient availability through nitrogen fixation and deep nutrient uptake.
7. Promoting efficient nutrtient cycling.

Classification of Problematic Soils: Problematic soils fall into two categories: physical and chemical.

1. Physical Problems:

- Fluffy paddy soil (low mechanical strength due to continuous rice cultivation).
- Sandy soils (resulting from weathering of granite, limestone, quartz; common in Haryana, U.P., Bihar).
- Hardpan soil (formation of compact layers due to illuviation of clay and metal oxides in red soils).
- Surface crusting (formation of a hardened soil layer due to iron and aluminum oxides in Alfisols).
- Peat and marshy soils (organic matter accumulation in humid regions, highly acidic, found in Kerala, Odisha, West Bengal).
- Waterlogged soil (excess moisture and poor aeration conditions).

2. Chemical Problems:

- **Acidic Soils:** Defined by pH below 7; extremely acidic soils have a pH between 4.0 and 4.75. Common in high rainfall zones due to base leaching and acidic parent materials (e.g., granite, sandstone). Found in Karnataka, M.P., Odisha, W.B., Kerala, Assam, Bihar.
- **Saline Soils:** Characterized by excessive soluble salts that affect plant growth.
- **Sodic Soils:** Contain high levels of exchangeable sodium, affecting soil structure.
- **Saline-Sodic Soils:** Exhibit traits of both saline and sodic soils, requiring sequential reclamation.

Adverse Effects of Acidic Soils on Plant Growth

1. Toxic solubility of elements like Al, Mn, and Fe at high acidity levels.
2. Inhibition of beneficial soil microorganisms.
3. Nutrient deficiencies, particularly of calcium (Ca) and potassium (K).

Reclamation Mechanisms for Saline Soils

1. Removing excess salts from the root zone.
2. Leaching with quality water and ensuring proper drainage.

3. Mulching to reduce salinity.
4. Adding organic matter to retain moisture and dilute salts.
5. Implementing green manuring and tree planting.
6. Selecting salt-tolerant species (*Prosopis juliflora, Tamarix articulata, Acacia nilotica*), and agricultural crops (*barley, sugar beet, cotton, wheat, rice, beans*).

Bioremediation through Agroforestry

- **Woody species for saline soils:** *Salvadora spp., Prosopis juliflora, Acacia nilotica, Parkinsonia aculeata, Butea monosperma, Terminalia arjuna, Salix spp., Dalbergia sissoo, Casuarina equisetifolia.*
- **Salt-tolerant grasses:** *Aeluropus lagopoides, Sporobolus helvolus, Cynodon dactylon, Brachiaria ramosa.*

Reclamation of Alkali Soils:

1. Ensuring adequate drainage to remove sodium accumulation.
2. Using salt-free irrigation water.
3. Adding organic matter and molasses.
4. Cultivating alkali-tolerant crops (*paddy, cotton, mustard, wheat, tomato, onion*).
5. Green manuring with *Dhaincha* for soil improvement.

Bioremediation through Agroforestry for Alkali Soils

- *Prosopis juliflora* and *Karnal grass* enhance soil conditions, allowing for the later introduction of high-quality fodder species such as *Berseem (Trifolium alexandricum), Senji (Melilotus parviflora), and Shaftal (Trifolium resupinatum).*

Saline-Sodic Soils and Their Reclamation: These soils exhibit traits of both salinity and sodicity. Initial reclamation targets salinity, followed by sodium mitigation. Cropping systems incorporating green manure or legumes are common.

- **Salt-tolerant crops:** *Rice, sugar beet, Dhaincha.*
- **Tree species for reclamation:** *Prosopis juliflora, Acacia nilotica, Prosopis chinensis.*

Bioremediation through Agroforestry for Saline-Sodic Soils

- *Acacia auriculiformis, Azadirachta indica, Casuarina equisetifolia, Dalbergia sissoo, Ailanthus excelsa, Prosopis cineraria, Acacia tortilis, Acacia nilotica.*

Conclusion

Multipurpose trees (MPTs) play a crucial role in enhancing sustainable land-use systems by providing ecological, economic, and social benefits. Their ability to serve multiple functions—such as soil fertility improvement, erosion control, carbon sequestration, fodder production, timber supply, and non-timber forest products—makes them indispensable in agroforestry, reforestation, and integrated farming systems. The careful selection of MPTs based on climatic adaptability, growth characteristics, and end-use requirements ensures their successful integration into diverse landscapes. By improving biodiversity, supporting livelihoods, and mitigating environmental degradation, multipurpose trees contribute significantly to climate resilience and food security. Therefore, promoting the cultivation of well-suited MPTs through research, policy support, and community engagement is essential for achieving sustainable land management and long-term environmental conservation. Their strategic incorporation into land-use systems can pave the way for a more productive and ecologically balanced future.

References

Assèdé, E. S., Sileshi, G. W., Chirwa, P. W., Orou, H., & Syampungani, S. (2024). Place and Roles of Trees in a Multifunctional Landscape: Trees and Environmental Services. In Trees in a Sub-Saharan Multi-functional Landscape: Research, Management, and Policy (pp. 41-58). Cham: Springer Nature Switzerland.

Behera, L., Nayak, M. R., Patel, D., Mehta, A., Sinha, S. K., & Gunaga, R. (2015). Agroforestry practices for physiological amelioration of salt aff ected soils. Journal of Plant Stress Physiology, 1, 13-18.

Borelli, S., Conigliaro, M., Quaglia, S., & Salbitano, F. (2018). Urban and Peri-urban agroforestry as multifunctional land use. In Agroforestry: anecdotal to modern science (pp. 705-724). Singapore: Springer Singapore.

Dagar, J. C., Singh, G., & Gupta, S. R. (2022). Sustainable land use systems for rehabilitation of highly sodic lands in the Indo-Gangetic plains of north-western India: synthesis of long-term field experiments. Journal of Soil Salinity and Water Quality, 14(1), 1-4.

Eshchanov, R., Kan, E., Khamzina, A., & Lamers, J. P. A. (2007). Farmer's knowledge and perceptions about multipurpose trees and tree intercropping systems in post-independence private farming systems in Uzbekistan ZEF Work Papers for Sustainable Development in Central Asia. ZEF Work Papers for Sustainable Development in Central Asia, (10).

Huxley, P. A. (1984). The Basis of Selection, Management, and Evaluation of Multipurpose Trees: An Overview.

Konijnendijk, C. C., & Gauthier, M. (2006). Urban forestry for multifunctional urban land use. Cities farming for the future: Agriculture for green and productive cities. R. van Veenhuizen, Ed. Rome: RUAF Foundation, 414-416.

Lelamo, L. L. (2021). A review on the indigenous multipurpose agroforestry tree species in Ethiopia: management, their productive and service roles and constraints. Heliyon, 7(9).

Nair, P. K. R. (1983, January). Multiple Land-Use and Agroforestry. In Ciba Foundation Symposium 97-Better Crops for Food (pp. 101-115). Chichester, UK: John Wiley & Sons, Ltd..

Nair, P. K. R., Fernandes, E. C. M., & Wambugu, P. N. (1984). Multipurpose leguminous trees and shrubs for agroforestry. Agroforestry systems, 2, 145-163.

Nair, P. R., Kumar, B. M., Nair, V. D., Nair, P. R., Kumar, B. M., & Nair, V. D. (2021). Multipurpose trees (MPTs) and other agroforestry species. An Introduction to Agroforestry: Four Decades of Scientific Developments, 281-351.

Okigbo, B. N. (1989). Role of multipurpose trees in compound farming in tropical Africa. In Alley farming in the humid and subhumid tropics: proceedings of an international workshop held at Ibadan, Nigeria, 10-14 Mar. 1986. IDRC, Ottawa, ON, CA.

Pandey, D. N. (2007). Multifunctional agroforestry systems in India. Current science, 455-463.

Ramesh, T., Manjaiah, K. M., Tomar, J. M. S., & Ngachan, S. V. (2013). Effect of multipurpose tree species on soil fertility and CO 2 efflux under hilly ecosystems of Northeast India. Agroforestry systems, 87, 1377-1388.

Saha, R., Tomar, J. M. S., & Ghosh, P. K. (2007). Evaluation and selection of multipurpose tree for improving soil hydro-physical behaviour under hilly eco-system of north east India. Agroforestry systems, 69(3), 239-247.

Singh, Y. P. (2017). Multifunctional agroforestry systems for bio-amelioration of salt-affected soils. Bioremediation of Salt Affected Soils: An Indian Perspective, 173-193.

Soulé, M., Kyereh, B., Kuyah, S., Tougiani, A., & Saadou, M. (2022). Azadirachta indica A. Juss. a multi-purpose tree as a leading species in carbon stocking in two Sahelian cities of Niger. Urban Ecosystems, 25(1), 51-64.

Torbert, J. L., & Burger, J. A. (2000). Forest land reclamation. Reclamation of drastically disturbed lands, 41, 371-398.

17

Potential of Agro Forestry Systems in Management of Problem Soils

Introduction

Agroforestry systems have the potential to make use of marginal and degraded lands through the soil improving effects of trees. Underlying all aspects of the role of agro forestry in maintenance of soil fertility is the fundamental proposition that trees improve soils. It would be useful to have guidelines on which properties of a tree or shrub species make it desirable for the point of view of soil fertility. This would help in identifying naturally occurring species and selecting trees for systems which have soil improvement as a specific objective.

Agroforestry is a sustainable land management system which increases the overall yield of the land, combines the production of crops (including tree crops) and forest plants and/or animals simultaneously or sequentially, on the same unit of land and applies management practices that are compatible with the cultural practices of the local population.

Nitrogen fixation and a high biomass production have been widely recognized as desirable. However, many properties are specific to particular objectives of systems in which the trees are used. Even species that are shunned for their competitive effects may have a role in certain designs. An example is the way in which Eucalyptus species with a high water uptake, which adversely affects yields in adjacent crops, have been employed to lower the water table and so reduce salinization

The agroforestry systems comprises Agrisilivicultural systems (improved fallow species in shifting agriculture (jhum), hedgerow intercropping (alley cropping), multispecies tree gardens, multipurpose trees/shrubs on farmlands, plantation and other crops, shade trees for commercial plantation crops, soil conservation hedges etc.), Agrisilvipastoral systems (tree- livestock-crop mix ground, home garden, woody hedgerows for browse, green manure, soil conservation etc., integrated production of crops, animals ,fuelwood, poles, etc.), Silvipostoral systems (multipurpose fodder trees on or around farmlands, living fences of fodder hedges and shrubs, trees and shrubs on pastures, integrated production of animals and wood products, etc.), aquaculture with trees, multipurpose tree lots, etc.

How do we know that trees improve soils

1. The soil that develops under natural forest and woodland is fertile. It is well structured, has a good water-holding capacity and has a store of nutrients bound up in the organic matter. Farmers know they will get a good crop by planting on cleared natural forest.
2. The cycles of carbon and nutrients under natural forest ecosystems are relatively closed, with much recycling and low inputs and outputs.
3. The practice of shifting cultivation demonstrated the power of trees to restore fertility lost during cropping.
4. Experience of reclamation forestry has demonstrated the power of trees to build up fertility on degraded land.

The properties which are likely to make a woody perennial suitable for soil fertility maintenance or improvement are:

- A high rate of production of leafy biomass.
- A dense network of fine roots, with a capacity for abundant mycorrhizal association.
- The existence of deep roots.
- A high rate of nitrogen fixation.
- A high and balanced nutrient content in the foliage; litter of high quality (high in nitrogen, low in lignin and polyphenols).
- An appreciable nutrient content in the root system
- Either rapid litter decay, where nutrient release is desired, or a moderate rate of litter decay, where maintenance of a soil cover is required.
- Absence of toxic substances in the litter or root residues
- For soil reclamation, a capacity to grow on poor soils.
- Absence of severe competitive effects with crops, particularly for water.
- Low invasiveness.
- Productive functions, or service functions other than soil improvement.

Not all of these properties are compatible: for example, litter of high quality is not likely to have a moderate rate of decay. The last property, the existence of productive functions, is not directly concerned with soils but is of the highest importance if the tree is to be effective in fertility maintenance. A species needs to be acceptable and desirable in agroforestry systems from other points of view, especially production.

A tree might have all the desirable properties above, but, if it is not planted and cared for, it will not be effective in improving soil fertility. The capacity of trees to maintain or improve soils is shown by the high fertility status and closed nutrient cycling under natural forest, the restoration of fertility under forest fallow in shifting cultivation, and the experience of reclamation forestry and agroforestry.

Soil transects frequently show higher organic matter and better soil physical properties under trees. Some species, most notably *Faidherbiaalbida*, regularly give higher crop yields beneath the tree canopy. Trees improve soil fertility by processes which.

- Increase additions to the soil
- Reduce losses from the soil
- Improve soil physical, chemical and biological conditions The most important sets of processes are those by which trees:
- Check runoff and soil erosion
- Maintain soil organic matter and physical properties
- Increase nutrient inputs, through nitrogen fixation and uptake from deep soil horizons
- Promote more closed nutrient cycling

Role of agroforestry in soil quality/heath

Compared to natural, a managed agricultural ecosystem has greater amounts of nutrient flowing in and out, less capacity for nutrient storage, and less nutrient recycling. The capacity of trees to maintain or improve soils is shown by the high fertility status and closed nutrient cycling under natural forest, the restoration of fertility under forest fallow in shifting cultivation, and the experience of reclamation forestry and agroforestry (Young, 2003). The processes by which trees maintain or improve soil fertility are given below:

1. Photosynthetic fixation of carbon and its transfer to the soil via litter and root decay,
2. Nitrogen fixation by all leguminous trees and in few non-leguminous species (e.g., Alder and Casuarinas),
3. Improved nutrient retrieval by tree roots, including through mycorrhiza and from lower horizon,
4. Providing favourable conditions for the input of nutrients from rainfall and dust

5. Control of erosion by combination of cover and barrier effect, especially the former,
6. Root uptake of nutrients that would otherwise have been lost by leaching,
7. Soils under trees have favourable structure and water holding capacity, through organic matter maintenance and root action,
8. Provision of a range of qualities of plant litter, woody, and herbaceous,
9. Growth promoting substances,
10. The potential through management of pruning and relative synchronizatoion of timing of release to nutrients from litter with demand for their uptake by crops, and
11. Effects of tree shading on microclimate

In agroforestry system nutrient addition takes place through leaf litter, pruning of woody compounds and atmospheric fixation. Some nutrients otherwise considered unavailable to crop because they are below the rooting zone of the annual crop, might be brought into the system from deeper layers in the soil with the the help of tree roots. Trees able to return nutrients through dead organic matter (leaf, branch, twig, fruits and flower) and thus helps in enrichment of top soil layer, available for the agriculture crops. Thus most important beneficial effect of the trees on the soil can include improvement of soil structure availability of nutrients. The objective, designing agroforestry system is to modify cycling in such a way as to make more efficient use of the nutrients whether these originate from natural removal process or from fertilizer. Trees in agroforestry system promote more closed nutrient cycling than pure agriculture systems.

Nutrient Recovery: Greater amount of nutrients that added to the soil, takes place through litterfall, trees translocate nutrients from deeper soil and deposit them on the soil surface via leaf sheading and other organic residues. The decomposition of organic matter residuces and its mineralization results release of nutrients to the soil. However, the amount of nutrients released in agroforestry systems will be much smaller than tree monoculture plantations

Biological Nitrogen Fixing: Agroforestry trees, particularly leguminous trees, enrich soil through biological nitrogen fixation, addition of organic matter and recycling of nutrients. Some trees such as Leucaena species, Acacia species and Alnus species has been reported to fix as much as 400-500 kg, 270 kg and 100-300 kg nitrogen per hectare per year respectively. The fixed nitrogen may benefit symbiotically to the crops growing in its association and helps in soil fertility improvement.

The amount of nitrogen added from the legumes or pruning of trees species taken up by the first crop is reported quite low and large portion is left in the soil organic matter indicating a long term nitrogen benefit than immediate. Different tree components viz., leaf, twigs, fruit and wood have different decomposition rates which helps to distribute the release of nutrient over time. Some important nitrogen fixing plant species are given in Table 1.

Biological nitrogen fixation takes place through symbioticand non-symbiotic means. Symbiotic fixation occurs through the association of plant roots with nitrogen-fixing microorganisms. Many legumes form an association with the bacteria Rhizobium while the symbionts of a few nonleguminous species belong to a genusofactinomycetes, Frankia. Non-symbiotic fixation is effected by free-living soil organisms, and can be a significant factor in natural ecosystems, which have relatively modest nitrogen requirements from outside systems (Nair, 1993)

Important N2 fixing plant species

Plant	**Botanical Name**	**Family**	**Nitrogen fixed (kg N/ha/yr)**
Black wattle	*Acacia mearnsii*	Mimosoideae	200
Beef wood, Saru	*Casurinaequisetif olia*	Casuarinaceae	60-110
Erythrina	*Erythrinapoeppigi ana*	Pipil[onaceae	60
Apple ring, Areca	*Gliricidiasepium*	Fabaceae	13
Inga	*Inga jincicuil*	Mimosoideae	34-50
Subabul	*Leucaenaleucoce phala*	Mimosoideae	100-500
Indian alder	*Alnusnepalensis*	Betulaceae	-
Horse bean	*Viciafaba*	Fabaceae	68-88

Nutrient Pumping: Tree root systems are involved in some favorable effects on soils such as carbon enrichment in soil through root turnover, the interception of leached nutrients, or the physical improvement of compact soil layers. Trees have deep and spreading roots and hence are capable of taking up nutrients and water from deeper soil layers usually where herbaceous crop roots cannot reach. This process of taking up nutrients from deeper soil profile and eventually depositing on the surface layers through litter-fall and other mechanisms is referred to as 'nutrient pumping' by trees. This process is mainly depends on characteristics of tree species and other soil, climatic and topographic factors.

Erosion Control: Loss of soil productivity and shortfall of food and cash crops are the immediate impacts of land degradation. India has 175 million ha degraded land. It is suffered from various problems of soil erosion and land degradation. These are major causes of land degradation and soil nutrient losses.

Amount of nutrient stored in above ground vegetation under different systems

Sl. No.	Species	DM (Ton/ha/yr)	Nutrient release (Kg/ha/yr)		
			N	P	K
1	Delbergia sissoo	4.2	67	3	16
2	Eucalyptus globules	8.5	58	5	48
3	Eucalyptus hybrid	4.6	22	9	7
4	Populus deltoids	4.5	25	16	26

DM = Dry matter

Soil: Tree species selected for agroforestry taking into account soil type

Desert soil: *Prosopis cineraria, P. chilensis, Acacia tortilis, A. senegal, A. nilotica, Salvadoraspp*

Recent alluvium: *Acacia catechu, Dalbergiasissoo, Bombaxceibaetc*

Saline-alkali soils: *Prosopisspp, Acacia nilotica, Azadirachtaindica, Ailanthus spp, Eucalyptus spp, Tamarixspp,Pongamiapinnata*

Coastal and deltaic alluvium: *Casuarinaequisetifolia, Cocusnucifera, Areca catechu, Avicenniaspp*

Red soils: *Tectonagrandis, Madhucaindica, Mangiferaindica, Dalbergiasissoo, Acacia nilotica, Leucaenaleucocephala, Azadirachtaindica, Eucalyptus hybrid, Pterocarpusmarsupium, Adina cardifolia Dendrocalamusstrictus*

Black cotton soils: *Acacia nilotica, A leucophloea, Tectonagrandis, Hardwickiabinnata, Adina cardifolia, Tamarandiusindica, Aeglemarmelos, Bauhinia spp, Dalbergialatifolia*

Laterite & lateric soils: *Tectonagrndis, Eucalyptus spp, Acacia auriculiformis, Azadirachtaindica, Tamarindusindica,Emblicaofficinalis*

Peaty and organic soil: *Syzygiumcuminii, Ficusglomerata, Bischofiajavanica, Lagerstromiaspeciosa, Glircidiasepium*

Hill soils: *Juglansregia, Alnusnitida, Toonaserrata, Cedrusdeodra, Quercusspp, Bombaxceiba etc.*

Conclusion

Agroforestry systems hold immense potential in the sustainable management of problem soils, offering a holistic approach to restoring soil health, enhancing productivity, and mitigating environmental degradation. By integrating trees, crops, and livestock, agroforestry improves soil structure, increases organic matter content, and enhances nutrient cycling, making it particularly effective for degraded, saline, acidic, or erosion-prone soils. The deep-rooting systems of trees help in breaking hardpans, improving water infiltration, and reducing surface runoff, thereby combating soil erosion and desertification. Additionally, nitrogen-fixing species replenish soil fertility, while litterfall from trees acts as a natural mulch, reducing evaporation and suppressing weeds. Agroforestry also contributes to carbon sequestration, helping mitigate climate change while providing economic benefits through diversified yields (timber, fruits, fodder, and crops). Systems like alley cropping, silvopasture, and windbreaks have proven successful in rehabilitating problem soils across different agro-climatic zones. However, the effectiveness of agroforestry depends on species selection, proper management practices, and local ecological conditions. Policy support, farmer education, and long-term research are essential to maximize its adoption and scalability. In conclusion, agroforestry represents a cost-effective, ecologically sound, and sustainable solution for managing problem soils, ensuring food security, environmental resilience, and improved livelihoods for farming communities. Its multifunctional benefits make it a key strategy in global efforts toward land restoration and climate-smart agriculture.

References

Assèdé, E. S., Sileshi, G. W., Chirwa, P. W., Orou, H., & Syampungani, S. (2024). Place and Roles of Trees in a Multifunctional Landscape: Trees and Environmental Services. In Trees in a Sub-Saharan Multi-functional Landscape: Research, Management, and Policy (pp. 41-58). Cham: Springer Nature Switzerland.

Behera, L., Nayak, M. R., Patel, D., Mehta, A., Sinha, S. K., & Gunaga, R. (2015). Agroforestry practices for physiological amelioration of salt aff ected soils. Journal of Plant Stress Physiology, 1, 13-18.

Borelli, S., Conigliaro, M., Quaglia, S., & Salbitano, F. (2018). Urban and Peri-urban agroforestry as multifunctional land use. In Agroforestry: anecdotal to modern science (pp. 705 721). Singapore: Springer Singapore.

Dagar, J. C., Singh, G., & Gupta, S. R. (2022). Sustainable land use systems for rehabilitation of highly sodic lands in the Indo-Gangetic plains of north-western India: synthesis of long-term field experiments. Journal of Soil Salinity and Water Quality, 14(1), 1-4.

Eshchanov, R., Kan, E., Khamzina, A., & Lamers, J. P. A. (2007). Farmer's knowledge and perceptions about multipurpose trees and tree intercropping systems in post-independence private farming systems in Uzbekistan ZEF Work Papers for Sustainable Development in Central Asia. ZEF Work Papers for Sustainable Development in Central Asia, (10).

Huxley, P. A. (1984). The Basis of Selection, Management, and Evaluation of Multipurpose Trees: An Overview.

Konijnendijk, C. C., & Gauthier, M. (2006). Urban forestry for multifunctional urban land use. Cities farming for the future: Agriculture for green and productive cities. R. van Veenhuizen, Ed. Rome: RUAF Foundation, 414-416.

Lelamo, L. L. (2021). A review on the indigenous multipurpose agroforestry tree species in Ethiopia: management, their productive and service roles and constraints. Heliyon, 7(9).

Nair, P. K. R. (1983, January). Multiple Land-Use and Agroforestry. In Ciba Foundation Symposium 97-Better Crops for Food (pp. 101-115). Chichester, UK: John Wiley & Sons, Ltd..

Nair, P. K. R., Fernandes, E. C. M., & Wambugu, P. N. (1984). Multipurpose leguminous trees and shrubs for agroforestry. Agroforestry systems, 2, 145-163.

Nair, P. R., Kumar, B. M., Nair, V. D., Nair, P. R., Kumar, B. M., & Nair, V. D. (2021). Multipurpose trees (MPTs) and other agroforestry species. An Introduction to Agroforestry: Four Decades of Scientific Developments, 281-351.

Okigbo, B. N. (1989). Role of multipurpose trees in compound farming in tropical Africa. In Alley farming in the humid and subhumid tropics: proceedings of an international workshop held at Ibadan, Nigeria, 10-14 Mar. 1986. IDRC, Ottawa, ON, CA.

Pandey, D. N. (2007). Multifunctional agroforestry systems in India. Current science, 455-463.

Ramesh, T., Manjaiah, K. M., Tomar, J. M. S., & Ngachan, S. V. (2013). Effect of multipurpose tree species on soil fertility and CO 2 efflux under hilly ecosystems of Northeast India. Agroforestry systems, 87, 1377-1388.

Saha, R., Tomar, J. M. S., & Ghosh, P. K. (2007). Evaluation and selection of multipurpose tree for improving soil hydro-physical behaviour under hilly eco-system of north east India. Agroforestry systems, 69(3), 239-247.

Singh, Y. P. (2017). Multifunctional agroforestry systems for bio-amelioration of salt-affected soils. Bioremediation of Salt Affected Soils: An Indian Perspective, 173-193.

Soulé, M., Kyereh, B., Kuyah, S., Tougiani, A., & Saadou, M. (2022). Azadirachta indica A. Juss. a multi-purpose tree as a leading species in carbon stocking in two Sahelian cities of Niger. Urban Ecosystems, 25(1), 51-64.

Torbert, J. L., & Burger, J. A. (2000). Forest land reclamation. Reclamation of drastically disturbed lands, 41, 371-398.